DRIVING STRATEGY
AND
INNOVATION WITH ANALYTICS

Samba Njong Joseph

Globally Available

Published in Nigeria, 2021
Emphaloz Publishing House

A catalogue record for this book is available at the National Library of Nigeria.

TABLE OF CONTENTS

FOREWORD

In today's rapidly evolving business landscape, the ability to adapt, innovate, and make informed decisions is no longer a luxury but a necessity. Data has become the cornerstone of successful enterprises, and analytics the process of transforming raw data into meaningful insights has emerged as the key to driving strategy and fostering innovation. As organisations across industries strive to stay competitive and navigate unprecedented challenges, the role of analytics in shaping the future of business strategy has never been more crucial. This book, Driving Strategy and Innovation with Analytics addresses a profound shift that is taking place in the way organisations operate. It is a shift away from traditional methods based on intuition and experience, and toward a data-driven approach that empowers leaders to anticipate trends, make smarter decisions, and innovate continuously. At the heart of this transformation is the ability to harness data not just for optimising operations, but to spark the creative, strategic thinking necessary for long-term success. For decades, businesses have relied on analytics to monitor performance and evaluate past actions. But the future of analytics goes beyond reflecting on the past. It is about using data to predict what's next and design strategies that are adaptive, resilient, and innovative.

In this book, we will explore how organisations are leveraging the power of analytics to drive strategy in real time, discover new opportunities for innovation, and cultivate a culture where data-driven decision-making becomes second nature. The convergence of technology, data, and business strategy has created a unique moment in time where organisations have more information at their fingertips than ever before. However, the ability to turn that information into actionable insights is where the true value lies. It's not about collecting massive amounts of data it's about understanding what the data is telling you and then acting. This is where analytics becomes not just a tool but a strategic partner in the innovation process. This book is both timely and essential. It provides a comprehensive guide for business leaders, entrepreneurs, and strategists who are seeking to harness the power of analytics to stay ahead of the curve. By demystifying the complexities of data and offering practical insights on how to integrate analytics into the core of business operations, Driving Strategy and Innovation with Analytics will equip readers with the knowledge they need to transform data into a strategic asset. In the chapters that follow, we will explore the ways in which data and analytics are driving innovation across industries.

From identifying new market opportunities to improving customer experiences, analytics is at the forefront of the most innovative ideas today. Readers will gain insight into how companies are using data to not only predict the future but to shape it creating strategies that are not only reactive but proactive. Moreover, this book will demonstrate that analytics is not just the domain of data scientists or IT departments. To fully realise the potential of analytics, organisations must cultivate a culture of data-driven thinking at every level. This means empowering employees with the tools and skills to use data in their daily decision-making processes. It means creating a collaborative environment where data is accessible, insights are shared, and innovation is encouraged. The businesses that succeed in the future will be those that embrace the full power of analytics not only to refine their strategies but to rethink the very nature of how they innovate. In this age of digital transformation, data is not just a resource; it is the foundation upon which future innovations will be built. Driving Strategy and Innovation with Analytics offers readers a roadmap to that future. It provides a step-by-step guide on how to incorporate analytics into every aspect of strategy development and innovation.

From setting clear objectives and defining key performance indicators (KPIs), to using advanced analytics tools and techniques for real-time decision-making, this book is a valuable resource for anyone looking to lead with data and inspire innovation. As we look ahead, the next wave of business innovation will be driven by those who can master the art of analytics. The tools and techniques are available; the challenge now is to apply them in a way that moves beyond data collection and into the realm of strategic foresight and creative disruption. I am confident that this book will inspire you to think differently about how you use data in your organisation. Whether you are a business leader seeking to improve performance or an entrepreneur aiming to disrupt your industry, this book will provide you with the insights, strategies, and tools to drive both innovation and growth through the power of analytics. As you embark on this journey, I encourage you to keep an open mind. Innovation, by its very nature, requires us to challenge conventional thinking, to take risks, and to embrace new ideas. With the guidance provided in this book, I have no doubt that you will be well-equipped to lead your organisation into a data-driven, innovative future.

INTRODUCTION

In an age where data is both abundant and essential, the ability to harness analytics for strategy and innovation has become the cornerstone of competitive advantage. Businesses today operate in environments characterised by rapid change, heightened competition, and shifting consumer demands. The organisations that excel are not necessarily those with the most resources, but those that can effectively leverage data to drive strategic decisions and foster continuous innovation. This book, Driving Strategy and Innovation with Analytics is about unlocking the potential of data to guide business strategy and fuel creativity. Data on its own is merely a resource, but when combined with sophisticated analytics, it becomes a powerful tool for identifying opportunities, predicting future trends, and creating innovative solutions that propel businesses forward. Whether you are a business leader, an entrepreneur, or a strategist, the ability to interpret and act on data insights will determine your success in navigating the complexities of today's marketplace. Strategy is about making informed choices choosing where to compete, how to allocate resources, and how to differentiate from competitors.

In the past, these decisions were often driven by intuition, experience, or limited historical data. While these approaches have their value, they lack the precision and foresight that analytics can provide. Today, data is available in real-time, offering insights into customer behaviour, market dynamics, and operational performance. Businesses that leverage analytics to inform their strategic decisions can move with greater agility, minimise risks, and capitalise on emerging opportunities. Analytics allows organisations to shift from reactive to proactive decision-making. Instead of waiting for changes in the market to occur, companies can use predictive analytics to anticipate these changes and adjust their strategies accordingly.

This forward-thinking approach enables businesses to not only survive in uncertain environments but thrive by staying ahead of competitors and responding to shifts in consumer preferences with speed and precision. Moreover, analytics plays a key role in optimising resource allocation. By analysing past performance and current data, companies can make better-informed decisions about where to invest their time, money, and efforts.

Whether it's deciding which products to develop, which markets to enter, or which segments to target, analytics ensures that these choices are based on data-backed insights rather than gut feelings or outdated information. While strategy focuses on positioning and resource allocation, innovation is about creating something new whether it's a product, service, or business model. Innovation is essential for businesses to stay relevant, meet evolving customer needs, and generate growth. However, innovation is often associated with risk. Businesses that rely solely on trial and error to innovate are at the mercy of unpredictable outcomes, and the cost of failure can be high. This is where analytics steps in as a game-changer. By using data to drive innovation, businesses can reduce the guesswork and uncertainty typically associated with the innovation process. Data provides insights into what customers truly want, where gaps exist in the market, and how products or services can be improved. It also allows businesses to test and iterate quickly, using real-time feedback to refine their offerings before fully launching them. Take, for example, companies that use customer data to inform product design. Through analytics, they can identify pain points, preferences, and patterns in customer behaviour that might not be obvious through traditional research methods.

This data-driven approach to innovation ensures that new products and services are aligned with market needs, increasing the likelihood of success. In addition, analytics empowers companies to pursue disruptive innovation breaking away from the status quo to create entirely new markets or business models. By analysing data from various industries, technologies, and consumer trends, businesses can identify opportunities that others might miss and position themselves as leaders in innovation. The promise of analytics extends far beyond the tools and techniques. For analytics to truly drive strategy and innovation, it must become ingrained in the organisational culture. This means fostering an environment where data-driven decision-making is not limited to a handful of analysts or executives but is practised at every level of the business. Creating a data-driven culture requires both top-down and bottom-up efforts. Leadership must champion the use of data and analytics in decision-making, demonstrating its value through clear communication and strategic initiatives. At the same time, employees need the skills and tools to work with data effectively.

This requires investment in data literacy programs and the democratisation of data access so that insights are not trapped in silos but are shared across teams to inform collaborative innovation efforts. When data-driven decision-making is part of the organisational DNA, businesses can move faster, respond more effectively to market changes, and innovate with confidence. It also encourages a mindset of continuous learning and improvement, where teams regularly revisit data to refine strategies and explore new possibilities for innovation. In Driving Strategy and Innovation with Analytics, we will explore how businesses can leverage data and analytics to transform their strategic thinking and unlock innovation. Each chapter will dive into specific aspects of this transformation, from using predictive analytics to identify market trends to employing data-driven techniques for optimising operations and fostering creativity. We will examine real-world case studies of companies that have successfully integrated analytics into their strategic and innovation processes. You will learn about the tools, techniques, and best practices that leading organisations use to stay ahead in a competitive landscape.

This book is designed to be both a practical guide and a source of inspiration. Whether you are looking to overhaul your organisation's approach to strategy, improve decision-making with data, or fuel innovation in your products and services, this book will provide you with the insights and frameworks you need to succeed in a data-driven world.

Let's begin this journey by exploring the power of analytics to not only guide your strategic decisions but also to spark the kind of innovation that drives sustainable growth

1

THE ROLE OF ANALYTICS IN STRATEGIC DECISION MAKING

In an era defined by rapid change, intense competition, and technological advancement, businesses that can harness the power of data to drive strategic decisions are better equipped to thrive. Traditional approaches to strategy, which often relied on intuition, past experiences, or historical trends, are no longer sufficient in today's fast-paced, data-driven world. The ability to collect, analyse, and act on data in real time has become an essential part of the strategic process, enabling organisations to remain agile, anticipate market shifts, and optimise resource allocation. The role of analytics in strategic decision-making goes beyond simply providing reports or dashboards. It involves embedding data-driven insights into every step of the decision-making process, from defining goals and identifying opportunities to assessing risks and measuring outcomes. Analytics transforms raw data into actionable insights, empowering businesses to make informed decisions that drive growth, improve efficiency, and foster innovation. In this chapter, we will explore how analytics serves as a critical tool for strategic decision-making. We will discuss the benefits of using analytics to inform strategy, examine real-world examples of data-driven decision-making in action, and highlight the key steps organisations can take to integrate analytics into their strategic processes.

Why Analytics Matters for Strategy

At its core, strategy is about making choices determining where to allocate resources, which markets to enter, what products to develop, and how to differentiate from competitors. The decisions involved in crafting a strategy are complex and often involve significant uncertainty. This is where analytics plays a crucial role. By leveraging data, businesses can reduce uncertainty, gain deeper insights into market trends, and make more informed decisions. One of the primary benefits of using analytics in strategic decision-making is the ability to predict future trends. Traditional strategic planning often relied on historical data and static forecasts, which provided a limited view of what the future might hold. In contrast, predictive analytics uses advanced statistical models and machine learning algorithms to analyse historical data and forecast future outcomes. This allows businesses to anticipate market shifts, customer behaviours, and competitive dynamics with greater accuracy. For example, a retail company might use predictive analytics to forecast customer demand for specific products during the holiday season. By analysing past sales data, current market trends, and external factors such as economic conditions, the company can predict which products are likely to be in high demand and adjust its inventory accordingly. This level of foresight helps the company avoid stockouts or overstock situations, leading to more efficient operations and higher customer satisfaction.

Another key advantage of analytics in strategy is its ability to optimise resource allocation. Businesses often face difficult choices when it comes to allocating resources whether it's deciding how much to invest in marketing, which products to prioritise, or where to expand operations. Analytics provides data-driven insights that help businesses allocate resources more effectively by identifying high-impact opportunities and minimising waste. For instance, a company might use analytics to assess the performance of its marketing campaigns across different channels, such as social media, email, and paid advertising. By analysing the return on investment (ROI) for each channel, the company can identify which channels are driving the most conversions and allocate more resources to those efforts. This data-driven approach ensures that the company is investing in the right areas to maximise its impact and achieve its strategic objectives.

Data-Driven Strategic Planning

Strategic planning is the process of defining an organisation's long-term goals and determining the actions needed to achieve those goals. Traditionally, strategic planning was often based on a combination of historical data, market research, and executive intuition. However, in a world where data is more accessible and abundant than ever before, strategic planning has evolved to become a more dynamic, data-driven process. Data-driven strategic planning involves using analytics to inform every stage of the planning process, from setting objectives and identifying opportunities to evaluating risks and monitoring progress. By leveraging data, businesses can create strategies that are more adaptive, resilient, and aligned with real-time market conditions. One of the first steps in data-driven strategic planning is to define clear objectives. Analytics can help organisations set more realistic and measurable goals by providing insights into past performance and current market trends. For example, if a company wants to increase its market share, it can use analytics to identify which customer segments offer the greatest growth potential and set specific, data-driven targets for each segment. Once objectives are defined, analytics can be used to identify opportunities for growth and innovation. For example, a business might analyse customer data to identify emerging trends or unmet needs in the market.

By understanding what customers want and where competitors are falling short, the business can develop new products or services that address these gaps, creating a competitive advantage. In addition to identifying opportunities, analytics plays a key role in risk assessment during the strategic planning process. Every strategic decision carries a certain level of risk, whether it's entering a new market, launching a new product, or investing in new technology. Analytics helps businesses assess these risks by analysing past outcomes, market volatility, and external factors such as economic conditions or regulatory changes. By understanding the potential risks associated with each decision, businesses can develop contingency plans and make more informed choices about which strategies to pursue. Finally, analytics is essential for monitoring progress and measuring the success of a strategic plan. Once a strategy is implemented, businesses need to track key performance indicators (KPIs) to ensure that they are on track to meet their goals. Analytics provides real-time insights into how well a strategy is performing, allowing businesses to adjust as needed. For example, if a company's sales data shows that a new product is underperforming in a particular market, the company can use analytics to investigate the cause and adjust its marketing or pricing strategy accordingly.

Real-World Examples of Analytics in Strategic Decision-Making

To understand the impact of analytics on strategic decision-making, it's helpful to look at real-world examples of companies that have successfully integrated data-driven insights into their strategic processes.

1. Netflix: Personalisation and Content Strategy

Netflix is a prime example of a company that uses analytics to drive its strategic decisions, particularly when it comes to content creation and personalisation. The streaming giant collects massive amounts of data on viewer behaviour, including what shows and movies users watch, how long they watch, and when they pause or stop a video. This data is then analysed to identify viewing patterns and preferences, which inform Netflix's content strategy. For instance, Netflix uses analytics to predict which types of shows or movies will be popular with specific audience segments. By analysing viewing trends, Netflix can make data-driven decisions about which original content to produce, and which licensed content to acquire. This approach not only helps Netflix stay ahead of competitors but also ensures that the company invests in content that resonates with its audience. In addition to content strategy, Netflix uses analytics to personalise the user experience. The platform's recommendation algorithm analyses each user's viewing history and suggests content that aligns with their preferences. This level of personalisation has been a key factor in Netflix's success, as it keeps users engaged and encourages them to spend more time on the platform.

2. Amazon: Optimising Operations with Data

Amazon is another company that has built its entire business strategy around data-driven decision-making. From inventory management to pricing strategies, Amazon uses analytics to optimise every aspect of its operations. One of the most notable examples of Amazon's use of analytics is its dynamic pricing model. Amazon uses data to track the prices of millions of products in real time, adjusting prices based on factors such as demand, competitor pricing, and inventory levels. This data-driven approach to pricing allows Amazon to remain competitive while maximising profitability. Amazon also uses predictive analytics to optimise its supply chain and inventory management. By analysing historical sales data, seasonal trends, and external factors such as weather patterns, Amazon can forecast demand for specific products and adjust its inventory levels accordingly. This level of precision helps Amazon avoid stockouts and overstock situations, leading to more efficient operations and higher customer satisfaction.

Integrating Analytics into Strategic Processes

For businesses looking to integrate analytics into their strategic decision-making processes, there are several key steps to consider:

Establish Clear Objectives: Before diving into data analysis, it's important to define what you hope to achieve with your strategy. Whether it's increasing market share, launching a new product, or improving operational efficiency, having clear objectives will help guide your analytics efforts and ensure that your decisions are aligned with your overall business goals.

Invest in Data Infrastructure: To leverage analytics effectively, businesses need the right data infrastructure in place. This includes investing in tools for data collection, storage, and analysis, as well as ensuring that data is accessible and up to date. Cloud-based data platforms, data lakes, and business intelligence tools are essential for managing large volumes of data and enabling real-time analysis.

Develop Data Literacy Across the Organisation: For analytics to truly drive strategy, it's important that employees at all levels of the organisation understand how to interpret and use data. Investing in data literacy programs and training can help ensure that decision-makers are comfortable working with data and are able to apply analytics to their day-to-day decision-making.

Build a Culture of Data-Driven Decision-Making: Finally, businesses need to create a culture where data-driven decision-making is encouraged and valued. This means empowering employees to use data in their decision-making processes, celebrating successes that result from data-driven insights, and promoting transparency in how decisions are made.

The role of analytics in strategic decision-making cannot be overstated. In a world where businesses must navigate complex, rapidly changing environments, the ability to leverage data to inform strategy is a critical competitive advantage. By using analytics to predict trends, optimise resource allocation, assess risks, and measure outcomes, businesses can make smarter, more informed decisions that drive growth and foster innovation. As we move forward in this book, we will explore how analytics can be applied to various aspects of strategy and innovation, providing a roadmap for organisations looking to harness the power of data to achieve long-term success.

2

USING ANALYTICS TO IDENTIFY OPPORTUNITIES FOR INNOVATION

Innovation is the lifeblood of growth and sustainability in any competitive business environment. It's what separates market leaders from followers, and it allows companies to adapt to changing circumstances, meet evolving customer needs, and create value in new ways. However, innovation doesn't occur in a vacuum. Successful innovation requires careful planning, insights into market demands, and an understanding of untapped opportunities. This is where analytics plays a critical role. Data-driven insights can help businesses identify opportunities for innovation that might otherwise go unnoticed. Whether it's developing new products, improving customer experiences, optimising processes, or entering new markets, analytics enables organisations to take a more strategic approach to innovation. By leveraging data to understand trends, identify gaps, and predict future opportunities, businesses can make informed decisions that drive meaningful innovation. In this chapter, we will explore how businesses can use analytics to discover opportunities for innovation across various areas of their operations. We'll cover key strategies for leveraging data to fuel innovation, examine real-world examples of companies using analytics to innovate, and provide a framework for integrating data-driven innovation into business practices.

The Link Between Analytics and Innovation

Innovation is often associated with creativity and new ideas, but in today's data-driven world, the most successful innovations are increasingly guided by analytics. While creativity is still essential, data provides the foundation for making informed decisions about which ideas to pursue, where to invest resources, and how to refine innovations to meet market needs. Analytics helps businesses move from abstract concepts of innovation to practical, actionable strategies. It enables organisations to quantify opportunities for innovation, providing clarity on market demand, customer pain points, and areas where competitors may be falling short. By identifying these gaps and opportunities, businesses can focus their innovation efforts on areas with the greatest potential for success. Furthermore, analytics allows businesses to test and iterate on their innovations in real time. By analysing customer feedback, market responses, and operational data, organisations can continuously refine their innovations, ensuring they stay aligned with evolving market conditions.

Using Customer Data to Drive Product Innovation

One of the most powerful applications of analytics in innovation is in product development. By analysing customer data such as purchasing behaviour, preferences, and feedback businesses can identify unmet needs and emerging trends, guiding the development of new products or the improvement of existing ones.

Understanding Customer Needs: Understanding what customers want is key to successful product innovation, and analytics provides a wealth of data that can help businesses gain this understanding. Customer relationship management (CRM) systems, social media platforms, and eCommerce websites generate massive amounts of data that reveal customer behaviours, preferences, and pain points. By analysing this data, businesses can gain insights into customer needs that may not be immediately apparent through traditional market research. For example, a company that sells fitness apparel might use data from its eCommerce site to identify which products customers are frequently searching for but not finding, indicating a gap in its product offerings. Similarly, the company could analyse customer reviews to identify common complaints or suggestions for improvement, such as requests for more sustainable materials or better fit. Through this analysis, the company can prioritise product innovations that address specific customer needs, increasing the likelihood that new products will resonate with its target audience. In this way, analytics allows businesses to take a more customer-centric approach to product innovation, ensuring that their efforts are aligned with what customers truly want.

Predicting Market Trends: In addition to understanding current customer needs, analytics can help businesses anticipate future market trends. By analysing historical data on customer behaviour, purchasing patterns, and external factors such as economic conditions, businesses can identify trends that are likely to shape future demand. This enables them to innovate proactively, developing products that meet emerging needs before competitors do. For example, a food and beverage company might use predictive analytics to forecast changes in consumer preferences related to health and wellness. By analysing data on dietary trends, social media conversations, and product sales, the company might identify a growing demand for plant-based protein products. Armed with this insight, the company can innovate by developing new plant-based offerings, positioning itself as a leader in a growing market segment. Predictive analytics allows businesses to stay ahead of the curve, ensuring that their innovations are not only relevant today but also positioned to meet future demand. This proactive approach to innovation can provide a significant competitive advantage, allowing businesses to capitalise on trends before they become mainstream.

Operational Analytics: Innovating Through Efficiency

Innovation isn't limited to products and services; it also applies to operational processes. By using analytics to optimise internal operations, businesses can innovate by finding new ways to improve efficiency, reduce costs, and enhance productivity. This type of innovation can have a profound impact on a company's bottom line, enabling it to deliver higher-quality products or services at a lower cost.

Process Optimisation: One of the most common applications of analytics in operational innovation is process optimisation. By analysing data on workflow processes, production cycles, and resource utilisation, businesses can identify inefficiencies and bottlenecks that are hindering performance. Armed with these insights, organisations can implement changes that streamline operations and improve overall efficiency. For example, a manufacturing company might use analytics to monitor the performance of its production lines. By analysing data on machine usage, downtime, and defect rates, the company can identify areas where production is slowing down or where quality issues are occurring. Based on this analysis, the company can adjust its processes such as scheduling maintenance during periods of low demand or automating certain tasks to improve efficiency and reduce waste. Process optimisation through analytics is not only about cutting costs; it's also about creating a more agile and responsive organisation. By continuously monitoring operational performance and making data-driven adjustments, businesses can innovate by developing processes that are more efficient, scalable, and adaptable to changing market conditions.

Supply Chain Innovation: Analytics also plays a crucial role in supply chain innovation. Supply chains are complex systems that involve multiple stakeholders, including suppliers, manufacturers, distributors, and customers. Managing these systems efficiently is critical to a company's success, and analytics provides the insights needed to optimise supply chain operations. By analysing data on supplier performance, inventory levels, transportation costs, and customer demand, businesses can identify opportunities to streamline their supply chains and reduce costs. For example, a company might use analytics to optimise its inventory management, ensuring that it has the right amount of stock on hand to meet customer demand without overstocking or understocking. This can reduce carrying costs, improve cash flow, and enhance customer satisfaction by ensuring products are available when needed. Additionally, analytics can help businesses mitigate supply chain risks. By analysing data on supplier reliability, geopolitical events, and market volatility, companies can identify potential disruptions in their supply chains and take proactive steps to address them. For example, a company might use analytics to identify alternative suppliers or adjust its production schedule in response to a potential shortage of raw materials.

Leveraging Competitive Intelligence for Innovation

Innovation doesn't occur in isolation; it happens within the context of a competitive landscape. To innovate successfully, businesses must not only focus on internal data but also keep an eye on what their competitors are doing. Competitive intelligence the process of gathering and analysing data on competitors can provide valuable insights that guide innovation.

Benchmarking Against Competitors: One-way businesses can use analytics to drive innovation is by benchmarking their performance against that of their competitors. By analysing data on market share, product features, pricing strategies, and customer reviews, businesses can identify areas where they are lagging and opportunities to differentiate themselves. For example, a software company might use competitive intelligence to compare its product features with those of its competitors. If the analysis reveals that competitors are offering certain features that customers value such as enhanced security or integration with other platforms the company can prioritise the development of similar features to stay competitive. In addition to product features, businesses can use competitive intelligence to benchmark their customer service and marketing strategies. By analysing customer satisfaction scores, response times, and social media engagement, businesses can identify areas where they can improve the customer experience or refine their marketing efforts to attract new customers.

Identifying White Space Opportunities: In addition to benchmarking against competitors, businesses can use analytics to identify white space opportunities areas of the market that are underserved or where competitors are not fully addressing customer needs. By analysing data on customer demographics, purchasing behaviours, and industry trends, businesses can uncover gaps in the market that present opportunities for innovation. For example, a company in the telecommunications industry might use competitive intelligence to identify that rural areas are underserved by high-speed internet providers. Based on this analysis, the company could innovate by developing a new service offering tailored to rural customers, filling a gap in the market and gaining a competitive edge. White space opportunities can exist in many areas, including product development, service delivery, and pricing models. By identifying these opportunities through analytics, businesses can create innovative solutions that meet unmet customer needs and differentiate themselves from competitors.

Creating a Culture of Data-Driven Innovation

For analytics to truly drive innovation, it must be embedded in the culture of the organisation. This means creating an environment where employees are encouraged to use data to explore new ideas, test hypotheses, and experiment with different approaches to solving problems. One way to foster a culture of data-driven innovation is by democratising data access. When employees at all levels of the organisation have access to data and the tools to analyse it, they are more likely to identify opportunities for innovation. This requires investing in user-friendly analytics platforms and training employees to work with data effectively. In addition, businesses should encourage cross-functional collaboration when it comes to innovation. Innovation often occurs at the intersection of different disciplines, and by bringing together teams from different departments such as marketing, product development, and operations businesses can generate new ideas and leverage data in creative ways. Finally, businesses should create an environment where experimentation is encouraged. Innovation involves taking risks, and not every idea will succeed. However, by using data to test and iterate on new ideas, businesses can reduce the risk associated with innovation and ensure that failures provide valuable learning opportunities. Analytics is a powerful tool for identifying opportunities for innovation. By leveraging data to understand customer needs, predict market trends, optimise operations, and analyse competitors, businesses can innovate more effectively and strategically. Whether it's developing new products, improving processes, or entering new markets, analytics provides the insights needed to guide innovation efforts and ensure they are aligned with business goals. As businesses continue to face increasing competition and rapidly changing market conditions, the ability to use analytics to drive innovation will be a key differentiator. In the next chapter, we will explore how businesses can use analytics to remain agile and adapt their strategies in response to real-time data, further enhancing their ability to innovate and succeed in a dynamic environment.

3

ANALYTICS FOR AGILE STRATEGY ADAPTATION.

In a business environment that is constantly evolving, the ability to adapt strategy in response to real-time data has become essential for long-term success. Traditional methods of strategic planning, which involved setting a long-term course and sticking to it, are increasingly ineffective in a world where market conditions, consumer behaviours, and technological advances can change at a rapid pace. To remain competitive, businesses must develop strategies that are flexible, adaptable, and responsive to new information. This is where analytics plays a crucial role. The rise of data-driven decision-making has given businesses the ability to continuously monitor their environment, assess the impact of their strategies, and adjust their course of action in real time. This agility allows organisations to pivot, when necessary, capitalise on emerging opportunities, and mitigate risks before they become significant challenges. In this chapter, we will explore how businesses can use analytics to develop agile strategies that respond to changing conditions. We'll discuss the key benefits of agile strategy adaptation, examine the role of real-time data and predictive analytics in enhancing agility, and provide a framework for integrating analytics into the process of strategy adjustment.

The Importance of Agility in Modern Business Strategy

Agility has become a buzzword in business, but its importance cannot be overstated. In the past, businesses could afford to set multi-year strategies based on historical trends and reasonably predictable market conditions. However, the advent of globalisation, digital transformation, and rapidly evolving customer expectations has created a more volatile and complex environment. As a result, companies must now be prepared to adjust their strategies frequently to stay relevant and competitive. An agile strategy is one that is flexible and adaptable, allowing businesses to respond to new information or changes in the market without abandoning their overarching goals. This approach enables organisations to seize emerging opportunities, avoid potential pitfalls, and stay ahead of competitors. But agility doesn't happen by chance it requires the right infrastructure, processes, and tools to support continuous monitoring, assessment, and adjustment of strategy. Analytics is the key enabler of agile strategy adaptation. By providing real-time insights into market conditions, customer behaviour, and operational performance, analytics empowers businesses to make informed decisions about when and how to adjust their strategies. Whether it's refining product offerings, shifting focus to new customer segments, or optimising marketing campaigns, analytics helps organisations respond more effectively to the dynamic nature of today's business environment.

Real-Time Data: The Foundation of Agile Decision-Making

The ability to adapt strategy in real time is only possible with access to real-time data. Real-time data refers to information that is collected, processed, and analysed as events occur, allowing businesses to make decisions based on the most up-to-date information available. This is a departure from traditional data analysis methods, which often relied on historical data that might not reflect current market conditions. Real-time data comes from a variety of sources, including customer interactions, social media activity, website traffic, financial transactions, and IoT (Internet of Things) devices. By continuously collecting and analysing this data, businesses can monitor key performance indicators (KPIs) in real time and quickly identify when something is going off track or when a new opportunity arises. For example, an eCommerce company might use real-time data to monitor website traffic during a product launch. If the data shows that customers are dropping off at a certain point in the checkout process, the company can quickly investigate the issue, adjust the user interface, and re-engage customers before losing sales. Similarly, if the company notices an unexpected surge in demand for a particular product, it can adjust its inventory management and marketing efforts to capitalise on the trend. Real-time analytics platforms, such as Google Analytics, Microsoft Power BI, and Tableau, make it easier for businesses to visualise and interpret real-time data. These platforms allow decision-makers to track key metrics, set up alerts for anomalies, and explore data trends in a user-friendly interface. By integrating real-time analytics into their strategic processes, businesses can respond more quickly to changing conditions and make data-driven adjustments to their strategies.

Predictive Analytics: Anticipating Changes Before They Happen

While real-time data provides valuable insights into what is happening now, predictive analytics takes it a step further by forecasting what is likely to happen in the future. Predictive analytics uses historical data, statistical models, and machine learning algorithms to predict future outcomes, allowing businesses to anticipate changes and adjust their strategies proactively. Predictive analytics is particularly valuable for identifying emerging trends, customer behaviour patterns, and potential risks. By understanding what is likely to happen in the future, businesses can make strategic decisions that position them for success. For example, a retail company might use predictive analytics to forecast customer demand for different products based on seasonal trends, economic conditions, and past purchasing behaviour. This allows the company to adjust its inventory and marketing strategies ahead of time, ensuring that it is well-prepared to meet customer needs. Predictive analytics can also help businesses mitigate risks. For instance, a financial services company might use predictive models to assess the likelihood of loan defaults based on customer data such as credit scores, payment histories, and economic indicators. By identifying high-risk customers, the company can adjust its lending strategy to minimise losses and protect its bottom line. One of the key advantages of predictive analytics is that it allows businesses to test different scenarios and evaluate the potential impact of various strategic decisions. For example, a company considering expanding into a new market might use predictive analytics to model different scenarios, such as how changes in customer demand, regulatory conditions, or competitive dynamics might affect its success. By exploring these scenarios, the company can make more informed decisions about whether and how to proceed with the expansion.

Adapting Strategy in Response to Market Shifts

In many industries, market conditions can change rapidly, and businesses must be prepared to adjust their strategies accordingly. Analytics provides businesses with the tools they need to monitor market shifts in real time and adapt their strategies to stay competitive. One of the keyways businesses can use analytics to adapt to market shifts is by tracking customer behaviour. Consumer preferences and behaviours can change quickly, especially in response to external factors such as economic conditions, technological advances, or social trends. By analysing customer data in real time, businesses can identify changes in buying patterns, preferences, or needs and adjust their strategies to meet these evolving demands. For example, during the COVID-19 pandemic, many businesses saw a sudden shift in consumer behaviour as customers moved away from in-store shopping to online shopping. Retailers that were able to quickly adapt to this shift by enhancing their eCommerce platforms, offering curb side pickup, and adjusting their marketing strategies were better positioned to weather the disruption. Analytics played a critical role in helping these businesses monitor the changing landscape and respond to new consumer demands. Analytics can also help businesses adapt their pricing strategies in response to market conditions. By analysing data on competitor pricing, customer demand, and production costs, businesses can make data-driven adjustments to their pricing models. For example, a company might use dynamic pricing algorithms to automatically adjust prices based on real-time data, ensuring that it remains competitive while maximising profitability.

Continuous Strategy Evaluation and Optimisation

An agile strategy is not a set-it-and-forget-it approach it requires continuous monitoring, evaluation, and optimisation. Businesses must regularly assess the performance of their strategies and make data-driven adjustments to ensure they remain aligned with their goals and the evolving market environment. Analytics provides the tools businesses need to evaluate their strategies in real time and make informed decisions about when and how to adjust them. Key performance indicators (KPIs) are an essential part of this process. KPIs provide measurable benchmarks that businesses can use to evaluate the success of their strategies. By tracking KPIs such as revenue growth, customer acquisition cost, or customer lifetime value, businesses can determine whether their strategies are delivering the desired results. However, tracking KPIs alone is not enough businesses must also analyse the underlying factors that are driving changes in these metrics. For example, if a company notices that its customer acquisition cost is increasing, it must investigate why this is happening. Analytics can help the company identify the root cause of the issue, whether it's a change in customer behaviour, a shift in the competitive landscape, or inefficiencies in its marketing efforts. Once the underlying issue is identified, businesses can use analytics to optimise their strategies. For example, if a company determines that its marketing campaigns are not resonating with a key customer segment, it can use data to refine its messaging, adjust its targeting, or explore new marketing channels. By continuously evaluating and optimising their strategies, businesses can ensure that they are staying aligned with their goals and remaining competitive in a changing environment.

Spotify's Data-Driven Strategy Adaptation

A real-world example of agile strategy adaptation through analytics can be seen in Spotify's approach to customer engagement and personalisation. Spotify is known for its use of data to create personalised user experiences, and the company continuously adapts its strategy based on real-time data and customer feedback. Spotify collects data on user behaviour, such as the songs and playlists users listen to, how often they engage with the app, and the types of content they prefer. The company uses this data to continuously refine its recommendation algorithms, ensuring that users are presented with music that aligns with their preferences. By analysing real-time data on user behaviour, Spotify can quickly identify when user engagement is dropping and adjust its strategy to re-engage users. For example, if Spotify notices that a particular user is listening to fewer songs or skipping more tracks than usual, it can use this data to adjust its recommendation engine, offering more relevant content to keep the user engaged. This level of personalisation helps Spotify retain users and improve their overall experience on the platform. In addition to adapting its recommendation algorithms, Spotify also uses data to optimise its marketing strategy. By analysing data on user demographics, listening habits, and engagement levels, Spotify can identify which customer segments are most likely to respond to specific marketing messages. This allows the company to create more targeted campaigns that resonate with its audience, leading to higher conversion rates and greater customer retention.

Implementing Analytics for Agile Strategy Adaptation

For businesses looking to implement analytics for agile strategy adaptation, there are several key steps to follow:

Invest in Real-Time Analytics Tools: To adapt strategies in real time, businesses need access to up-to-date information. Investing in real-time analytics platforms that can process and visualise data as it is collected is critical for making informed decisions quickly.

Set Clear KPIs and Metrics: Establish measurable KPIs that align with your strategic goals. These metrics will serve as benchmarks for evaluating the success of your strategies and determining when adjustments are needed.

Use Predictive Analytics to Anticipate Changes: Predictive analytics allows businesses to anticipate future trends and challenges before they occur. By integrating predictive models into your strategic planning process, you can proactively adjust your strategies to stay ahead of the competition.

Foster a Culture of Agility: Agile strategy adaptation requires a culture that encourages flexibility, experimentation, and data-driven decision-making. Ensure that employees at all levels of the organisation have access to data and are empowered to make strategic adjustments as needed.

Continuously Monitor and Optimise: Agile strategy is an ongoing process. Continuously monitor the performance of your strategies, analyse the data, and adjust as needed to ensure that your organisation remains competitive and responsive to changing conditions.

In a world where business conditions can change rapidly and unpredictably, agility is no longer a competitive advantage it is a necessity. By using analytics to develop agile strategies, businesses can respond more quickly to market shifts, optimise their operations in real time, and stay ahead of competitors. Whether it's through real-time data monitoring, predictive analytics, or continuous strategy evaluation, analytics provides the tools needed to adapt and thrive in a dynamic environment. In the next chapter, we will explore how fostering a culture of innovation with data can drive continuous improvement and create an environment where creativity and analytics work hand in hand to fuel business growth.

4

FOSTERING A CULTURE OF INNOVATION WITH DATA.

Innovation is essential for business growth and sustainability in a fast-paced, competitive marketplace. However, for innovation to flourish, it must be embedded in the company culture, not just relegated to specific teams or projects. When data is harnessed effectively, it can become the foundation for a culture of innovation, driving continuous improvement and creative problem-solving across the organisation. In this chapter, we will explore how businesses can foster a culture of innovation by leveraging data as a key enabler. We'll discuss the importance of democratising data access, promoting data literacy, encouraging experimentation, and supporting cross-functional collaboration. By embedding data into every aspect of the innovation process, businesses can unlock new opportunities, optimise their operations, and create a sustained competitive advantage.

The Role of Data in Driving Innovation

Data has become a critical asset for businesses seeking to innovate. It offers the ability to uncover insights about customer preferences, market trends, operational inefficiencies, and potential growth opportunities.

However, data on its own does not drive innovation. It is the strategic application of data combined with a company-wide commitment to innovation that yields results. Data-driven innovation involves more than just analysing historical data to inform decision-making; it requires using data to identify emerging trends, forecast future opportunities, and fuel the creative process. When employees are empowered to use data to experiment, solve problems, and make decisions, they become more proactive in driving innovation. Moreover, data can reduce the risks typically associated with innovation. Innovation often involves venturing into uncharted territory, and the outcomes can be uncertain. However, data provides businesses with the ability to test hypotheses, validate ideas, and refine solutions before fully committing to them. This reduces the likelihood of failure and increases the chances of success by ensuring that innovations are aligned with market needs and backed by evidence.

Democratising Data Access

One of the key components of fostering a culture of innovation with data is democratising data access. In many organisations, data is siloed within specific departments, such as IT, marketing, or finance, making it difficult for employees outside of these teams to access and use it. This lack of access can stifle innovation by limiting the ability of employees to identify opportunities, test ideas, or make data-driven decisions. To create a culture of data-driven innovation, businesses must make data accessible to employees across all levels and functions of the organisation. This can be achieved using self-service analytics platforms that allow non-technical users to access and analyse data without needing assistance from data scientists or IT teams. Tools such as Power BI, Tableau, and Google Data Studio are designed with user-friendly interfaces that empower employees to generate reports, visualise data, and extract insights independently.

By democratising data access, businesses enable employees from diverse backgrounds and skill sets to contribute to the innovation process. For example, a product development team might use data on customer feedback to inform the design of a new product, while a marketing team could analyse data on customer behaviour to refine a campaign strategy. When data is accessible to everyone, it becomes a tool for collaboration and creativity, enabling teams to approach problems from different perspectives and generate innovative solutions. However, democratising data access is not just about providing tools it also requires a shift in organisational mindset. Leadership must promote the idea that data is a shared resource and that every employee has a role to play in using it to drive innovation. This cultural shift encourages employees to take ownership of data and feel empowered to use it in their decision-making processes.

Promoting Data Literacy

While democratising data access is important, it is equally essential to ensure that employees have the skills to use data effectively. This is where data literacy comes into play. Data literacy refers to the ability to read, understand, analyse, and communicate data. In a data-driven organisation, data literacy is a fundamental skill that all employees should possess, regardless of their role or function. Promoting data literacy is a critical step in fostering a culture of innovation. When employees are comfortable working with data, they are more likely to use it to inform their decisions, test new ideas, and explore innovative solutions. Conversely, if employees lack the skills to interpret data, they may be hesitant to engage with it or rely on outdated methods of decision-making. To improve data literacy, businesses can implement training programs that teach employees how to work with data and analytics tools. These programs should cover the basics of data analysis, such as understanding data types, interpreting visualisations, and identifying trends.

More advanced training might include topics such as predictive analytics, data modelling, and hypothesis testing. It's also important to create opportunities for employees to practise using data in their day-to-day work. This can be done by encouraging employees to experiment with data in low-stakes environments, such as using analytics tools to analyse internal processes or test new ideas. By fostering a hands-on approach to data literacy, businesses can help employees build confidence in their ability to use data to drive innovation. Additionally, leadership should emphasise the importance of data literacy by embedding it into the company's values and objectives. This can be done by recognising and rewarding employees who use data to innovate or make data-driven decisions. When data literacy is seen as a core competency, employees are more likely to invest the time and effort needed to develop their skills.

Encouraging Experimentation and Risk-Taking

Innovation inherently involves risk, as it requires businesses to step outside of their comfort zones and explore new ideas. However, for innovation to thrive, employees must feel empowered to take risks and experiment with new approaches. Data plays a crucial role in mitigating the risks associated with innovation by providing a foundation for testing, learning, and iterating. One-way businesses can encourage experimentation is by adopting a test-and-learn approach to innovation. This approach involves using data to test new ideas on a small scale before fully implementing them. By running experiments, businesses can gather data on how a new product, service, or process performs in real-world conditions. If the data shows promising results, the innovation can be scaled up; if the data reveals challenges or shortcomings, the business can iterate and refine the idea. For example, a retail company might test a new pricing strategy by rolling it out in a limited number of stores. By analysing sales data, customer feedback, and competitor pricing, the company can assess the impact of the strategy before deciding whether to implement its companywide.

This data-driven experimentation reduces the risk of failure while allowing the company to innovate in a controlled, evidence-based manner. Encouraging experimentation also requires creating a culture where failure is seen as a learning opportunity rather than a setback. In many organisations, fear of failure can stifle innovation, as employees may be reluctant to pursue new ideas if they are unsure of the outcome. However, when businesses use data to measure and learn from failures, they can turn setbacks into valuable insights that inform future innovation efforts. Leadership plays a key role in fostering a culture of experimentation by promoting a growth mindset. This mindset encourages employees to view challenges and failures as opportunities for growth and improvement. By framing failure as a natural part of the innovation process, leaders can create an environment where employees feel safe to experiment, take risks, and learn from their experiences.

Supporting Cross-Functional Collaboration

Innovation often occurs at the intersection of different disciplines, and data can act as a catalyst for cross-functional collaboration. When employees from different departments and functions come together to share insights, they bring diverse perspectives that can spark new ideas and solutions. Data enables cross-functional teams to collaborate more effectively by providing a common language and framework for decision-making. For example, a marketing team, a product development team, and an operations team might work together to launch a new product. By analysing customer data, market trends, and supply chain performance, the teams can align their efforts and make data-driven decisions that ensure the product's success. One way to promote cross-functional collaboration is by creating data-sharing platforms that allow teams to access and analyse data from different parts of the organisation. For example, a shared dashboard might display key metrics related to sales, customer satisfaction, and operational performance. By providing access to this data, businesses can break down silos and enable teams to work together more effectively. In addition to creating data-sharing platforms, businesses should encourage a culture of open communication and collaboration. This can be done by organising cross-functional meetings, innovation workshops, or hackathons where employees from different departments come together to brainstorm ideas and solve problems using data. By fostering a collaborative environment, businesses can leverage the collective expertise of their employees to drive innovation.

Data-Driven Leadership: Setting the Tone for Innovation

Leadership is critical to the success of any innovation effort, and in a data-driven organisation, leaders must set the tone for how data is used to drive innovation. Data-driven leadership involves more than just promoting the use of data it requires leaders to model data-driven decision-making, encourage experimentation, and create an environment where innovation can thrive. Leaders can demonstrate their commitment to data-driven innovation by making data a key part of the strategic decision-making process. This means using data to inform decisions at every level, from setting long-term goals to making day-to-day operational choices. When employees see that leadership values data-driven decision-making, they are more likely to adopt the same approach in their own work. In addition to modelling data-driven behaviour, leaders should actively support innovation initiatives by providing the resources and tools needed for success. This includes investing in data infrastructure, analytics platforms, and training programs that empower employees to use data effectively. Leaders should also provide opportunities for employees to experiment with new ideas and reward those who take risks and drive innovation.

Finally, data-driven leaders must foster a culture of transparency and accountability. This involves being open about how data is used to make decisions and holding teams accountable for using data to measure the impact of their innovations. By promoting transparency and accountability, leaders can create an environment where data is seen as a valuable tool for driving continuous improvement and innovation. Fostering a culture of innovation with data requires more than just investing in analytics tools it involves creating an environment where data-driven decision-making is encouraged, experimentation is supported, and collaboration is promoted. By democratising data access, promoting data literacy, encouraging risk-taking, and supporting cross-functional collaboration, businesses can unlock the full potential of data to drive innovation. As we continue to explore the role of data in business strategy and innovation, the next chapter will focus on how predictive and prescriptive analytics can be used to drive innovation by identifying opportunities, optimising processes, and guiding decision-making.

5

PREDICTIVE AND PRESCRIPTIVE ANALYTICS IN DRIVING INNOVATION.

Innovation is often seen as a creative endeavour, fuelled by bold ideas and intuitive leaps. However, in today's data-driven world, the most successful innovations are increasingly powered by analytics specifically, predictive and prescriptive analytics. These advanced forms of analytics enable businesses to anticipate future trends, optimise their processes, and make more informed decisions that drive innovation. Predictive and prescriptive analytics take data beyond the realm of historical insights, enabling organisations to forecast what is likely to happen and recommend actions that align with strategic goals. In this chapter, we will explore how predictive and prescriptive analytics can be used to fuel innovation. We'll examine the differences between the two approaches, highlight real-world applications, and provide a framework for integrating predictive and prescriptive analytics into the innovation process.

Understanding Predictive Analytics

Predictive analytics is the practice of using historical data, statistical algorithms, and machine learning techniques to predict future outcomes. It is commonly used to forecast trends, identify risks, and estimate future behaviours or events. By analysing patterns in past data, predictive analytics can help businesses anticipate what might happen under various conditions, enabling them to make proactive decisions that drive innovation. At its core, predictive analytics answers the question, "What is likely to happen?" This forward-looking approach allows businesses to stay ahead of the curve by anticipating market shifts, customer preferences, and operational challenges before they fully materialise.

How Predictive Analytics Drives Innovation

Predictive analytics has become an essential tool for innovation across multiple industries. By using data to forecast future events, businesses can identify opportunities for new products, services, or business models before their competitors. Here are a few ways predictive analytics can drive innovation:

Product Development and Customisation: Predictive analytics is invaluable in developing new products or enhancing existing ones. By analysing customer data, businesses can identify patterns in preferences, behaviours, and feedback that reveal unmet needs or emerging demands. For instance, in the consumer electronics industry, companies use predictive models to anticipate the features and functionalities customers will demand in the next generation of smartphones. This allows them to stay ahead of trends and deliver products that meet future customer expectations. In addition to predicting demand, predictive analytics can be used to personalise products. By analysing individual customer data, businesses can offer personalised recommendations or tailor products to meet specific needs. This level of customisation not only enhances customer satisfaction but also drives brand loyalty and differentiation.

Customer behaviour Forecasting: Understanding customer behaviour is critical for businesses looking to innovate in customer experience and service delivery. Predictive analytics allows businesses to forecast how customers will respond to changes in the market, such as new products, price fluctuations, or marketing campaigns. For example, retailers use predictive analytics to forecast how customers will respond to seasonal promotions, allowing them to optimise inventory and marketing strategies in advance. Predictive analytics can also be used to identify customers at risk of churning. By analysing customer interactions, purchase histories, and engagement patterns, businesses can predict which customers are likely to leave and take proactive measures to retain them. This not only helps preserve customer loyalty but also opens opportunities for innovation in customer service and engagement strategies.

Operational Efficiency and Risk Mitigation: Predictive analytics can be used to optimise operations and improve efficiency. For example, in manufacturing, predictive analytics is used to forecast equipment failures, allowing companies to perform maintenance before a breakdown occurs. This predictive maintenance reduces downtime and operational costs, enabling businesses to operate more efficiently. Risk mitigation is another area where predictive analytics shines. By analysing data on market conditions, regulatory changes, or economic indicators, businesses can forecast potential risks and develop strategies to mitigate them. This proactive approach to risk management allows businesses to innovate by adapting to changing conditions before they become major challenges.

Understanding Prescriptive Analytics

While predictive analytics forecasts future outcomes, prescriptive analytics goes a step further by providing recommendations on how to achieve desired outcomes. Prescriptive analytics uses data, algorithms, and optimisation techniques to suggest specific actions that will lead to the best possible results. It answers the question, "What should we do?" rather than just "What is likely to happen?" Prescriptive analytics is particularly valuable for businesses looking to innovate because it provides actionable insights that guide decision-making. Rather than relying on intuition or guesswork, businesses can use prescriptive analytics to optimise their strategies, processes, and operations based on data-driven recommendations.

How Prescriptive Analytics Drives Innovation

Prescriptive analytics can be a powerful tool for driving innovation in areas such as decision-making, process optimisation, and resource allocation. Here are several ways in which prescriptive analytics contributes to innovation:

Optimising Business Processes: Prescriptive analytics can be used to optimise business processes by recommending the best course of action based on data-driven insights. For example, in supply chain management, prescriptive analytics can analyse data on inventory levels, demand forecasts, and supplier performance to recommend the most efficient way to allocate resources. This helps businesses reduce costs, minimise waste, and improve delivery times, all of which contribute to innovation in supply chain operations. In addition to supply chain management, prescriptive analytics can be used to optimise marketing campaigns, pricing strategies, and production schedules. By continuously analysing data and recommending adjustments, prescriptive analytics helps businesses innovate by making their operations more agile and responsive to changing conditions.

Improving Decision-Making with Actionable Insights: One of the key benefits of prescriptive analytics is that it provides decision-makers with actionable insights that guide their choices. This is particularly valuable in industries where decision-making is complex and high stakes, such as finance, healthcare, and manufacturing. For example, in the healthcare industry, prescriptive analytics is used to recommend personalised treatment plans based on patient data. By analysing data on patient outcomes, medical history, and treatment options, prescriptive analytics can suggest the most effective course of action, improving patient care and reducing costs. In the finance industry, prescriptive analytics can be used to recommend investment strategies based on market trends, risk tolerance, and financial goals. By providing specific recommendations on how to allocate resources, prescriptive analytics helps investors optimise their portfolios and maximise returns.

Supporting Strategic Innovation: Prescriptive analytics is a valuable tool for supporting strategic innovation by providing businesses with recommendations on how to achieve their long-term goals. For example, a company looking to expand into a new market might use prescriptive analytics to analyse data on customer preferences, competitor performance, and market conditions. Based on this analysis, the company can receive recommendations on the best way to enter the market, such as which products to prioritise, how to position itself against competitors, and what pricing strategy to adopt. This data-driven approach to strategic innovation helps businesses make informed decisions about where to invest their resources and how to differentiate themselves from competitors. By optimising their strategies based on prescriptive analytics, businesses can reduce the risks associated with innovation and increase their chances of success.

Combining Predictive and Prescriptive Analytics for Maximum Impact

While predictive and prescriptive analytics are valuable on their own, they are even more powerful when used together. Predictive analytics provides businesses with insights into what is likely to happen, while prescriptive analytics offers recommendations on what actions to take in response to those predictions. By combining these two approaches, businesses can create a comprehensive, data-driven innovation strategy that is both forward-looking and actionable. For example, a retail company might use predictive analytics to forecast a surge in demand for a particular product based on seasonal trends and customer data. Armed with this insight, the company can then use prescriptive analytics to determine the best way to respond, such as adjusting inventory levels, launching a targeted marketing campaign, or offering discounts to capitalise on the increased demand. In this way, predictive and prescriptive analytics work together to drive innovation by enabling businesses to anticipate changes and take proactive steps to optimise their strategies. This combination of foresight and action is critical for staying competitive in today's fast-paced business environment.

Real-World Applications of Predictive and Prescriptive Analytics

To illustrate how predictive and prescriptive analytics can drive innovation, let's look at a few real-world examples:

Amazon: Predictive Analytics for Customer Experience - Amazon uses predictive analytics to enhance the customer experience and drive innovation in eCommerce. By analysing customer behaviour, purchase history, and browsing patterns, Amazon's predictive algorithms can recommend products that customers are likely to be interested in. This personalised shopping experience has been a key factor in Amazon's success, allowing the company to innovate continuously by offering customers the products they want before they even realise, they need them. In addition to predicting customer preferences, Amazon uses predictive analytics to optimise its inventory management and supply chain operations. By forecasting demand for different products, Amazon can ensure that its warehouses are stocked with the right inventory at the right time, reducing costs and improving delivery times.

General Electric: Prescriptive Analytics for Industrial Efficiency - General Electric (GE) is a pioneer in using prescriptive analytics to optimise industrial operations. GE's Predix platform uses prescriptive analytics to monitor the performance of industrial equipment, such as jet engines, wind turbines, and power plants. By analysing data from sensors embedded in the equipment, Predix can recommend maintenance schedules, operational adjustments, and resource allocations that optimise performance and reduce downtime. This use of prescriptive analytics has allowed GE to drive innovation in the industrial sector by improving efficiency, reducing costs, and extending the lifespan of its equipment. By providing actionable recommendations based on real-time data, GE's prescriptive analytics platform helps businesses make smarter, more informed decisions about how to operate their assets.

Netflix: Predictive and Prescriptive Analytics for Content Strategy - Netflix is known for its use of predictive analytics to forecast viewer preferences and recommend content to its users. By analysing data on viewing habits, ratings, and preferences, Netflix can predict which shows and movies are likely to resonate with specific audience segments. This allows Netflix to innovate continuously by offering personalised recommendations that keep users engaged and subscribed. In addition to predictive analytics, Netflix uses prescriptive analytics to guide its content creation strategy. By analysing data on viewer engagement, production costs, and market trends, Netflix can make data-driven decisions about which shows and movies to produce or licence. This prescriptive approach to content strategy has allowed Netflix to stay ahead of competitors and deliver original content that consistently meets audience demand.

Implementing Predictive and Prescriptive Analytics in Innovation

For businesses looking to integrate predictive and prescriptive analytics into their innovation processes, the following steps can help ensure success:

Define Clear Objectives: Before implementing predictive or prescriptive analytics, businesses must define their innovation objectives. Whether it's developing a new product, entering a new market, or optimising operations, having clear goals will guide the analytics process and ensure that the insights generated are aligned with the company's strategic priorities.

Invest in the Right Tools and Technologies: Predictive and prescriptive analytics require sophisticated tools and technologies, such as machine learning algorithms, optimisation models, and advanced analytics platforms. Businesses should invest in the infrastructure needed to collect, process, and analyse data in real time.

Ensure Data Quality: The accuracy and effectiveness of predictive and prescriptive analytics depend on the quality of the data used. Businesses must ensure that their data is clean, accurate, and up to date. Implementing data governance practices, such as data validation and cleansing, will help improve the quality of the insights generated by analytics.

Foster a Data-Driven Culture: Predictive and prescriptive analytics are only effective if they are integrated into the decision-making process. Businesses should foster a culture of data-driven innovation by encouraging employees to use analytics to inform their decisions and experiment with new ideas.

Monitor and Iterate: Predictive and prescriptive analytics are not one-time processes. Businesses should continuously monitor the performance of their analytics models, test new hypotheses, and iterate on their strategies to ensure that they remain aligned with changing market conditions and customer needs.

Predictive and prescriptive analytics are powerful tools for driving innovation by providing businesses with the insights and recommendations they need to make informed decisions. By forecasting future trends and recommending specific actions, these advanced forms of analytics enable businesses to anticipate changes, optimise operations, and innovate more effectively. When used together, predictive and prescriptive analytics provide a comprehensive, data-driven approach to innovation that allows businesses to stay ahead of competitors and create sustained value. In the next chapter, we will explore how businesses can use analytics to drive innovation in product and service development, leveraging data to create offerings that meet evolving customer needs and market demands.

6

DATA-DRIVEN PRODUCT AND SERVICE INNOVATION.

Innovation is essential for any business aiming to remain competitive and relevant in today's rapidly evolving market. But the days of relying solely on intuition, creativity, or conventional market research to drive product and service innovation are gone. In a world where data is ubiquitous, companies that effectively leverage analytics have a distinct advantage in creating innovative products and services that meet the needs of their customers while staying ahead of competitors. Data-driven innovation allows businesses to base their product and service development decisions on hard evidence customer preferences, usage patterns, emerging market trends, and even predictive forecasts. The integration of analytics into the innovation process transforms traditional research and development (R&D) efforts by aligning innovation with real-time data, minimising the risk of failure, and maximising the potential for success. In this chapter, we'll explore how companies can leverage data to drive product and service innovation. We'll cover how data helps businesses identify opportunities for new products, improve existing services, personalise customer experiences, and anticipate future trends. We'll also provide real-world examples of companies that have successfully used data to fuel innovation and offer a step-by-step framework for integrating data into the product and service innovation process.

The Role of Data in Identifying Opportunities for Innovation

The foundation of any successful product or service innovation is identifying unmet customer needs or gaps in the market. Traditionally, businesses relied on surveys, focus groups, or market analysis reports to understand these needs. While these methods are still valuable, they can be slow and often only providing a snapshot of a moment in time. Data-driven approaches, on the other hand, provide real-time, continuous insights that enable businesses to respond quickly to changes in customer preferences and market dynamics.

Analysing Customer Behaviour to Identify Gaps

One of the most powerful ways to identify opportunities for innovation is by analysing customer behaviour data. By understanding how customers interact with existing products and services, businesses can uncover pain points, preferences, and emerging needs that might not be immediately obvious through traditional methods. For example, usage data from a mobile app might reveal that users frequently abandon the app after encountering a certain feature, indicating a need for improvement or redesign. In eCommerce, analysing purchase behaviour can reveal gaps in a company's product offerings. If data shows that customers frequently search for a product that isn't available or frequently abandon their carts when certain items are out of stock, this signals an opportunity to introduce new products or optimise the product mix. Customer reviews, social media interactions, and customer service inquiries are other valuable data sources. These channels can provide qualitative data that highlights what customers like or dislike about a product or service, allowing businesses to address specific pain points or enhance popular features. By systematically analysing this feedback, companies can gain deeper insights into customer needs and use this information to guide product and service development.

Using Predictive Analytics to Anticipate Future Needs

In addition to identifying current gaps in the market, businesses can use predictive analytics to forecast future trends and customer needs. Predictive analytics uses historical data, machine learning algorithms, and statistical models to predict future behaviours, preferences, or market shifts. This enables companies to innovate proactively rather than reactively, developing products or services that align with future demand. For instance, in the fashion industry, predictive analytics is used to anticipate style trends based on historical sales data, social media conversations, and external factors such as seasonality or cultural events. Retailers can then design and produce clothing lines that meet anticipated customer preferences, ensuring they stay ahead of competitors. Similarly, in the tech industry, companies use predictive analytics to forecast which features or functionalities customers will demand in the next generation of products. By analysing data on customer preferences, product usage, and technology adoption rates, tech companies can prioritise their R&D efforts and invest in innovations that are likely to succeed in the market.

Personalising Products and Services Through Data

One of the most significant trends in product and service innovation is personalisation. Today's customers expect products and services to be tailored to their individual preferences, needs, and behaviours. Data-driven personalisation allows businesses to deliver highly customised experiences that increase customer satisfaction and foster loyalty.

Personalisation in Product Development

In product development, data can be used to create personalised offerings based on individual customer preferences. For example, in the automotive industry, manufacturers use data on customer preferences, driving habits, and lifestyle to offer personalised features or design options. A customer who frequently drives long distances might be offered a vehicle with enhanced fuel efficiency or advanced navigation features, while a customer who values luxury might be offered premium interior options. Another example of personalisation in product development is in the health and wellness industry, where companies use data to create customised products based on an individual's health data, dietary habits, or genetic profile. For instance, personalised vitamin subscriptions are growing in popularity, with companies using customer-provided health information to recommend personalised vitamin regimens tailored to everyone's needs. Data-driven personalisation is not only beneficial for customers but also for businesses, as it enables companies to differentiate themselves from competitors and build deeper relationships with their customers. By offering products that are uniquely suited to each customer's needs, businesses can increase customer retention and lifetime value.

Personalising Services with Data

In addition to personalising products, data is being used to personalise services in industries such as healthcare, finance, and hospitality. In healthcare, for instance, data-driven personalisation allows providers to offer personalised treatment plans based on an individual's medical history, lifestyle, and genetic makeup. This approach, known as precision medicine, improves patient outcomes by tailoring treatments to the specific needs and characteristics of each patient. In the financial industry, companies use data to offer personalised investment advice, credit offers, or financial planning services. By analysing data on an individual's financial behaviour, spending habits, and risk tolerance, financial institutions can provide personalised recommendations that help customers achieve their financial goals. In the hospitality industry, hotels and resorts use data to personalise the guest experience. By analysing data on past stays, preferences, and feedback, hotels can offer personalised services such as room preferences, dining options, or activity recommendations. This level of personalisation enhances the guest experience and encourages repeat business.

Iterative Innovation: Using Data to Continuously Improve Products and Services

Innovation is not a one-time event it's an ongoing process that requires continuous refinement and improvement. Data enables businesses to adopt an iterative approach to product and service development, where they continuously collect feedback, test new ideas, and make data-driven adjustments.

A/B Testing and Experimentation - One of the most effective ways to use data for iterative innovation is through A/B testing. A/B testing involves comparing two versions of a product or service to determine which one performs better based on specific metrics, such as user engagement, conversion rates, or customer satisfaction. For example, a software company might use A/B testing to compare two versions of a user interface design. By analysing data on user interactions, the company can determine which design leads to better user engagement or satisfaction. Based on these insights, the company can make data-driven decisions about which design to implement. A/B testing can be applied to various aspects of product or service development, including pricing strategies, marketing campaigns, or feature enhancements. By continuously testing and iterating, businesses can optimise their offerings and ensure that they meet the evolving needs of their customers.

Real-Time Data and Agile Product Development In industries where customer preferences or market conditions can change rapidly, businesses must be able to adapt quickly. Real-time data allows businesses to monitor customer behaviour, product performance, and market trends in real-time, enabling them to make quick adjustments to their products or services. For example, in the software industry, companies use real-time data to monitor how users interact with their applications. If data shows that a new feature is not being used as expected, the company can quickly investigate the issue, adjust, and release an updated version of the software. This agile approach to product development allows businesses to respond to customer feedback and market changes in real-time, ensuring that their products remain relevant and competitive. Real-time data is also valuable for optimising supply chain operations. For instance, retailers use real-time inventory data to ensure that products are available when and where customers need them. If a product is selling faster than expected, the retailer can use real-time data to adjust its supply chain and ensure that additional inventory is shipped to the appropriate locations.

Case Studies: Data-Driven Innovation in Action

To illustrate how data drives product and service innovation, let's look at a few real-world examples:

Spotify: Personalising the Music Experience with Data - Spotify has built its entire business model around data-driven personalisation. By analysing data on users' listening habits, preferences, and engagement, Spotify's recommendation algorithm delivers personalised playlists and song suggestions that keep users engaged and satisfied. The company's Discover Weekly feature, which recommends new music to users based on their listening history, is a prime example of how data can be used to deliver highly personalised experiences. Spotify also uses data to innovate its product offerings. For example, the company analyses data on how users engage with different features such as playlists, podcasts, or social sharing to continuously improve the user experience and introduce new features that align with user behaviour. This data-driven approach to innovation has helped Spotify maintain its position as a leader in the music streaming industry.

Tesla: Data-Driven Innovation in Automotive Design - Tesla is known for its use of data to drive innovation in the automotive industry. Tesla vehicles are equipped with sensors that collect vast amounts of data on driving behaviour, vehicle performance, and environmental conditions. This data is used not only to improve the design and functionality of Tesla's cars but also to inform the development of autonomous driving technology. By continuously collecting and analysing data from its vehicles, Tesla can make software updates in real time, improving vehicle performance and introducing new features without requiring customers to purchase new models. This data-driven approach to innovation has enabled Tesla to stay at the forefront of the automotive industry, particularly in the development of electric and autonomous vehicles.

Netflix: Innovating Content Strategy with Data - Netflix uses data to drive its content strategy and personalise the viewing experience for its users. By analysing data on viewing habits, user ratings, and engagement, Netflix can predict which shows and movies will resonate with its audience and invest in original content that is likely to succeed. The company's data-driven approach to content development has led to the creation of hit shows like Stranger Things and House of Cards, both of which were developed based on insights from viewer data. In addition to content development, Netflix uses data to personalise the user experience. The company's recommendation algorithm suggests shows and movies that align with each user's preferences, increasing engagement and reducing churn. This level of personalisation has been a key factor in Netflix's success and continued growth in the streaming industry.

Implementing Data-Driven Innovation in Product and Service Development

For businesses looking to leverage data to drive product and service innovation, the following steps can help ensure success:

Collect Comprehensive Data: The first step in data-driven innovation is collecting comprehensive data from multiple sources, including customer interactions, market trends, and operational performance. The more data businesses have, the better they can understand customer needs and identify opportunities for innovation.

Invest in Analytics Tools: To analyse and interpret the data, businesses need to invest in advanced analytics tools that can process large datasets, generate insights, and visualise trends. Tools such as predictive analytics platforms, customer relationship management (CRM) systems, and business intelligence (BI) software are essential for turning data into actionable insights.

Foster Cross-Functional Collaboration: Data-driven innovation requires collaboration between different teams, including product development, marketing, and operations. By fostering a culture of collaboration, businesses can ensure that data is shared across teams and used to inform all aspects of product and service development.

Continuously Monitor and Adjust: Innovation is an ongoing process, and businesses must continuously monitor the performance of their products and services to ensure they remain aligned with customer needs. By using real-time data to track product usage, customer feedback, and market trends, businesses can make data-driven adjustments and iterate on their innovations.

Data-driven innovation is transforming the way businesses develop products and services, allowing them to create offerings that are tailored to customer needs, personalised to individual preferences, and continuously improved through iterative testing and real-time adjustments. By leveraging data throughout the innovation process, businesses can minimise the risks associated with product development, enhance the customer experience, and stay ahead of competitors. In the next chapter, we will explore how analytics can be used to gain a competitive advantage by identifying trends before competitors, optimising operations, and differentiating in the marketplace.

7

ANALYTICS FOR COMPETITIVE ADVANTAGE IN INNOVATION.

In the increasingly competitive global marketplace, businesses are constantly searching for ways to differentiate themselves, stay ahead of competitors, and create sustained value for their customers. The traditional methods of gaining competitive advantage such as offering lower prices or providing better customer service are still relevant, but they are no longer sufficient on their own. Today, the ability to harness data through analytics has emerged as a key driver of competitive advantage, especially in the realm of innovation. By leveraging analytics, businesses can identify trends before their competitors, anticipate customer needs, optimise internal processes, and create innovative products or services that set them apart in the marketplace. Analytics provides organisations with the insights needed to not only keep pace with competitors but to leap ahead by offering solutions that are uniquely tailored to market demands. In this chapter, we'll explore how businesses can use analytics to gain a competitive advantage in innovation. We'll examine how data-driven insights help organisations identify trends, optimise operations, tailor offerings to customer segments, and outmanoeuvre competitors. We'll also provide real-world examples of companies that have successfully used analytics to drive innovation and gain a strategic edge.

The Role of Analytics in Identifying Trends and Opportunities

In a fast-moving market, the ability to identify emerging trends before competitors is a significant competitive advantage. Analytics allows businesses to sift through vast amounts of data to uncover patterns, behaviours, and trends that can shape future market conditions. By analysing customer behaviour, industry shifts, and broader macroeconomic indicators, businesses can spot opportunities for innovation before they become obvious to others.

Analysing Customer Data for Market Trends - One of the most valuable sources of insight for gaining a competitive advantage is customer data. By analysing data on customer preferences, purchasing behaviours, and engagement patterns, businesses can identify emerging trends that signal shifts in the market. For example, analysing purchase data from an eCommerce platform might reveal that customers are increasingly interested in sustainable or eco-friendly products. Armed with this insight, a business can introduce a new line of sustainable products ahead of competitors, positioning itself as a leader in this growing market. Customer data also helps businesses understand which features or services are driving customer satisfaction and loyalty. By analysing this data, businesses can prioritise product improvements or new features that align with customer needs, giving them a competitive edge in delivering a superior customer experience.

Monitoring Industry Data for Competitive Intelligence - Beyond customer data, businesses can use analytics to monitor industry-wide data for competitive intelligence. This involves analysing data from sources such as industry reports, competitor financial filings, and market research studies to identify trends that might impact the competitive landscape. For example, if an industry report shows that demand for a particular type of product is growing rapidly in certain geographic regions, a company can use this insight to expand its operations into those regions before competitors do.

Competitive intelligence can also be gleaned from social media, news sources, and even patent filings. By tracking conversations, product launches, and industry announcements, businesses can stay informed about what competitors are doing and adjust their strategies accordingly. The ability to respond to competitors' moves in real time, or even anticipate them, is a powerful way to maintain a competitive advantage.

Predictive Analytics for Future Trends - In addition to analysing current data, businesses can use predictive analytics to forecast future trends and opportunities. Predictive analytics uses historical data and machine learning algorithms to identify patterns and predict what is likely to happen in the future. This forward-looking approach enables businesses to anticipate changes in customer preferences, market conditions, or competitive dynamics before they occur. For example, a retailer might use predictive analytics to forecast which product categories will see increased demand during the holiday season based on historical sales data, social media trends, and macroeconomic indicators. By preparing for this surge in demand, the retailer can stock inventory, accordingly, optimise its marketing strategy, and ensure that it is positioned to capture a larger share of the market. Predictive analytics is especially valuable for first-mover advantages, where being the first to market with a new product or service can lead to significant gains. By predicting customer needs or industry shifts before they become widely apparent, businesses can introduce innovations that meet emerging demand, securing a competitive edge before rivals have a chance to react.

Tailoring Offerings to Specific Customer Segments

Another way analytics provides a competitive advantage is by enabling businesses to segment their customer base and tailor their offerings to meet the specific needs of each segment. In today's marketplace, personalisation is a key differentiator, and customers expect businesses to understand and cater to their individual preferences. Analytics allows businesses to analyse customer data at a granular level, identifying unique segments within their customer base and developing targeted products, services, or marketing campaigns for each segment.

Customer Segmentation and Personalisation - Customer segmentation involves dividing a company's customer base into distinct groups based on characteristics such as demographics, purchasing behaviour, or preferences. By using analytics to perform customer segmentation, businesses can gain a deeper understanding of the specific needs and desires of each group, allowing them to offer personalised experiences that resonate with customers. For example, a travel company might segment its customer base into groups such as business travellers, families, and adventure seekers. By analysing data on booking history, travel preferences, and spending patterns, the company can develop customised travel packages or marketing campaigns that appeal to the unique needs of each group. Business travellers might receive offers for quick, convenient business trips, while families might be targeted with vacation packages that include family-friendly activities and accommodations. Personalisation not only improves customer satisfaction but also increases customer loyalty, as customers are more likely to return to businesses that offer tailored experiences. This level of personalisation gives businesses a competitive advantage by making their offerings more relevant and appealing to individual customers compared to competitors who take a one-size-fits-all approach.

Dynamic Pricing Models - In addition to tailoring products and services, analytics can be used to implement dynamic pricing models that optimise pricing based on real-time data. Dynamic pricing allows businesses to adjust prices in response to changes in demand, inventory levels, competitor pricing, or customer behaviour. This flexible approach to pricing enables businesses to maximise revenue and gain a competitive edge by offering the right price at the right time. For example, ride-sharing companies like Uber and Lyft use dynamic pricing to adjust fares based on factors such as demand, traffic conditions, and time of day. During peak hours or in high-demand areas, prices increase to reflect the scarcity of available drivers. This data-driven approach to pricing ensures that companies can meet customer demand while maximising profitability. Dynamic pricing models can also be applied to industries such as retail, hospitality, and eCommerce. Retailers might use dynamic pricing to offer discounts on slow-moving inventory or increase prices on high-demand items during peak shopping seasons. By using data to optimise pricing in real time, businesses can improve their competitiveness and respond more effectively to market conditions.

Optimising Operations for Competitive Advantage

Innovation is not limited to products and services it also extends to internal operations. Analytics provides businesses with the tools they need to optimise their operations and gain a competitive advantage by improving efficiency, reducing costs, and enhancing performance. By analysing data on processes, supply chains, and resource allocation, businesses can identify inefficiencies and implement data-driven solutions that streamline operations and improve their bottom line.

Supply Chain Optimisation - One of the most impactful ways to gain a competitive advantage through analytics is by optimising the supply chain. Supply chains are complex networks that involve multiple stakeholders, including suppliers, manufacturers, distributors, and retailers. Managing these networks efficiently is critical to ensuring that products are delivered to customers on time and at the right cost. Analytics allows businesses to analyse data on inventory levels, supplier performance, transportation costs, and demand forecasts to optimise supply chain operations. For example, by using predictive analytics to forecast demand for specific products, businesses can adjust their inventory levels to avoid stockouts or overstock situations. This level of precision helps businesses reduce carrying costs, improve cash flow, and enhance customer satisfaction by ensuring that products are available when and where they are needed. Additionally, businesses can use prescriptive analytics to optimise their logistics and distribution networks. By analysing data on transportation routes, fuel costs, and delivery times, businesses can identify the most efficient ways to transport goods, reducing shipping costs and improving delivery speed. This level of optimisation not only reduces operational costs but also provides a competitive edge by enabling faster and more reliable delivery compared to competitors.

Process Innovation and Automation - In addition to optimising the supply chain, businesses can use analytics to innovate internal processes and implement automation. By analysing data on workflow processes, production cycles, and resource utilisation, businesses can identify inefficiencies and bottlenecks that hinder performance. Armed with these insights, businesses can implement automation technologies that streamline operations and reduce the need for manual intervention. For example, in manufacturing, analytics can be used to monitor equipment performance and identify maintenance needs before breakdowns occur. By implementing predictive maintenance strategies, businesses can reduce downtime, extend the lifespan of equipment, and improve overall operational efficiency. This proactive approach to maintenance not only reduces costs but also improves production reliability, giving businesses a competitive advantage by ensuring that operations run smoothly. In industries such as finance, healthcare, and retail, analytics-driven automation is used to improve customer service and reduce manual processes. For instance, chatbots powered by AI and natural language processing (NLP) can handle routine customer inquiries, freeing up human agents to focus on more complex issues. Similarly, automated workflows can be used to process transactions, schedule appointments, or manage inventory, improving efficiency and reducing the risk of human error.

Analytics-Driven Competitive Advantage in Action

To illustrate how analytics can be used to gain a competitive advantage in innovation, let's look at a few real-world examples:

Walmart: Supply Chain Optimisation through Data Analytics - Walmart is renowned for its use of data analytics to optimise its supply chain operations and maintain its competitive edge in the retail industry. By analysing data on inventory levels, supplier performance, and customer demand, Walmart has implemented a just-in-time inventory system that ensures products are delivered to stores exactly when they are needed. This data-driven approach to inventory management reduces carrying costs and minimises the risk of stockouts, allowing Walmart to offer lower prices than its competitors while maintaining high levels of customer satisfaction. Walmart also uses predictive analytics to forecast demand for products based on factors such as seasonality, economic conditions, and regional preferences. This level of precision allows Walmart to optimise its supply chain, ensuring that products are available in the right quantities and at the right time.

Coca-Cola: Personalisation through Data-Driven Marketing - Coca-Cola uses data analytics to personalise its marketing campaigns and engage with customers in meaningful ways. By analysing data on customer preferences, social media interactions, and purchase history, Coca-Cola can create targeted marketing campaigns that resonate with specific customer segments. For example, Coca-Cola's Share a Coke campaign, which featured personalised labels with customers' names, was driven by data insights that showed a growing trend toward personalisation. In addition to personalised marketing, Coca-Cola uses data to optimise its distribution networks and improve operational efficiency. By analysing data on sales patterns, delivery routes, and inventory levels, Coca-Cola can ensure that its products are always available in the right locations, reducing costs and improving customer satisfaction.

UPS: Using Predictive Analytics for Operational Efficiency - UPS has long been a leader in using predictive analytics to optimise its logistics and delivery operations. By analysing data on delivery routes, traffic conditions, and fuel costs, UPS developed its ORION (On-Road Integrated Optimisation and Navigation) system, which uses predictive algorithms to determine the most efficient routes for its delivery drivers. ORION helps UPS reduce fuel consumption, lower emissions, and improve delivery times, giving the company a significant competitive advantage in the logistics industry. UPS also uses predictive analytics to forecast package volume during peak shipping seasons, allowing the company to adjust its workforce and resources accordingly. By optimising its operations through data, UPS can provide faster, more reliable delivery services than its competitors, ensuring customer satisfaction and loyalty.

Implementing Analytics for Competitive Advantage

For businesses looking to gain a competitive advantage through analytics, the following steps can help ensure success:

Invest in Data Infrastructure: To leverage analytics effectively, businesses need to invest in the infrastructure required to collect, process, and analyse data. This includes data storage systems, analytics platforms, and business intelligence tools.

Set Clear Objectives: Competitive advantage through analytics starts with clear objectives. Businesses must define what they hope to achieve, whether it's improving customer satisfaction, reducing operational costs, or entering new markets.

Develop a Data-Driven Culture: Analytics-driven competitive advantage requires a culture where data is valued and used to inform decision-making at all levels of the organisation. This includes fostering data literacy among employees and promoting collaboration between data teams and business units.

Monitor and Iterate: Gaining a competitive advantage is an ongoing process. Businesses should continuously monitor the performance of their analytics initiatives and adjust as needed to stay ahead of competitors. In today's hyper-competitive market, analytics provides businesses with the insights needed to gain a strategic edge through innovation. By using data to identify trends, personalise offerings, optimise operations, and outmanoeuvre competitors, businesses can create a sustained competitive advantage that drives growth and success. In the next chapter, we will explore how businesses can use analytics to foster innovation in internal operations and processes, driving efficiency and continuous improvement.

8

INNOVATION IN OPERATIONS AND PROCESSES THROUGH ANALYTICS.

While product and service innovation often grab the headlines, the most significant opportunities for innovation frequently lie within a company's internal operations and processes. Operational innovation driven by data and analytics can deliver transformative results, improving efficiency, reducing costs, enhancing quality, and ultimately boosting a company's competitive advantage. Analytics provides businesses with the insights needed to streamline workflows, optimise resource allocation, reduce waste, and implement automation that leads to more agile, efficient, and scalable operations. In today's fast-paced business environment, operational excellence is a key driver of sustained growth. Companies that use analytics to innovate their operations are better positioned to adapt to changing market conditions, meet customer demands more effectively, and ensure their long-term profitability. In this chapter, we'll explore how businesses can leverage data and analytics to drive innovation in their internal operations and processes. We'll cover specific applications such as process optimisation, supply chain management, predictive maintenance, and workforce productivity. We'll also look at real-world examples of companies that have successfully used data to transform their operations.

The Importance of Operational Innovation

Operational innovation focuses on transforming the way a company conducts its internal activities to achieve greater efficiency, quality, or speed. It involves rethinking and redesigning processes to eliminate inefficiencies, reduce costs, and improve overall performance. While product and service innovation may capture the imagination of customers, operational innovation is often the foundation that allows businesses to deliver those products and services more effectively. Historically, operational improvements have been driven by methodologies such as Lean, Six Sigma, and Total Quality Management (TQM), all of which aim to eliminate waste and enhance quality. However, the advent of advanced analytics has opened new opportunities for operational innovation by providing deeper insights into every aspect of a company's processes. Analytics allows businesses to move beyond basic process improvements by using real-time data and predictive insights to optimise every stage of operations. Whether it's optimising a manufacturing line, improving inventory management, or enhancing customer service, data-driven insights enable businesses to make informed decisions that lead to significant operational improvements.

Process Optimisation through Data-Driven Insights

Process optimisation is one of the most direct ways to use analytics to drive operational innovation. Every business, regardless of its size or industry, relies on processes whether it's manufacturing, order fulfilment, customer service, or supply chain management. The efficiency and effectiveness of these processes have a direct impact on a company's profitability, customer satisfaction, and overall competitiveness. By analysing data on process performance, resource utilisation, and workflow efficiency, businesses can identify bottlenecks, inefficiencies, and areas for improvement. This allows companies to make data-driven adjustments that streamline operations, reduce costs, and improve output quality.

Workflow Efficiency - One of the most common applications of process optimisation is in improving workflow efficiency. Workflow inefficiencies can manifest in many ways, such as excessive waiting times between tasks, underutilisation of resources, or redundant activities that don't add value. By analysing data on how tasks flow through a process, businesses can identify and eliminate these inefficiencies. For example, in a manufacturing environment, a company might use analytics to track the movement of materials through the production line. If the data shows that certain machines or workstations are frequently idle or that materials are being delayed at specific points in the process, the company can adjust improve flow and reduce downtime. This might involve redistributing work among different stations, automating certain tasks, or implementing new scheduling practices to ensure that resources are being utilised more effectively. Similarly, in service industries, such as call centres or customer support, analytics can be used to optimise workflow efficiency by analysing data on call handling times, customer wait times, and agent performance. By identifying patterns in this data, businesses can redesign their workflows to reduce wait times, improve response times, and enhance overall service quality.

Resource Allocation - Another critical aspect of process optimisation is resource allocation. Resources whether they are people, equipment, or materials are often one of the most significant costs for businesses. Ensuring that these resources are allocated effectively is essential for optimising operational performance. Data analytics enables businesses to monitor resource usage in real time and adjust as needed. For instance, in a retail environment, a company might use data on customer foot traffic and sales patterns to allocate staff more effectively during peak hours. By ensuring that the right number of employees are on hand when demand is highest, the company can improve customer service while minimising labour costs. In manufacturing, data analytics can be used to monitor equipment usage and optimise scheduling. By analysing data on machine performance, production schedules, and order volumes, businesses can ensure that equipment is being utilised at maximum capacity without causing overuse or burnout. This not only improves efficiency but also extends the lifespan of expensive machinery, reducing maintenance and replacement costs.

Supply Chain Optimisation through Predictive Analytics

Supply chain management is another area where analytics-driven innovation can have a transformative impact. Supply chains are inherently complex, involving multiple stakeholders, locations, and processes. Any inefficiencies or disruptions in the supply chain can lead to delays, increased costs, and lost revenue. By using data analytics to optimise supply chain operations, businesses can enhance efficiency, reduce risks, and improve overall performance.

Demand Forecasting - One of the keyways businesses can use analytics to optimise their supply chain is through demand forecasting. Predictive analytics enables companies to forecast customer demand based on historical sales data, market trends, seasonality, and external factors such as economic conditions or weather patterns. Accurate demand forecasting allows businesses to plan their inventory and production schedules more effectively, ensuring that they have the right products in the right quantities at the right time. For example, in the retail industry, a company might use predictive analytics to forecast demand for specific products during the holiday season. By analysing data on past holiday sales, current market trends, and customer preferences, the company can adjust its inventory levels and ensure that it has enough stock to meet customer demand without overstocking. In addition to improving inventory management, demand forecasting can help businesses optimise their supply chain by reducing the risk of stockouts or excess inventory. By aligning production and procurement with accurate demand forecasts, businesses can avoid costly overproduction or underproduction, leading to more efficient use of resources and improved profitability.

Supplier Performance Management - Another critical aspect of supply chain optimisation is supplier performance management. Businesses rely on a network of suppliers to provide the raw materials, components, or services they need to produce their products. The performance of these suppliers such as their reliability, lead times, and cost-effectiveness directly impacts the efficiency of the supply chain. Data analytics allows businesses to monitor and evaluate supplier performance based on key metrics such as on-time delivery rates, defect rates, and responsiveness. By analysing this data, companies can identify high-performing suppliers and negotiate better terms, while also identifying underperforming suppliers that may need to be replaced or renegotiated with. For example, a manufacturing company might use analytics to track the performance of its suppliers over time. If the data reveals that certain suppliers consistently fail to meet delivery deadlines or provide substandard materials, the company can take corrective action, such as diversifying its supplier base or renegotiating contracts. This proactive approach to supplier management ensures that the supply chain remains efficient and reliable, reducing the risk of disruptions.

Predictive Maintenance and Asset Management

In industries such as manufacturing, transportation, and utilities, equipment downtime can have a significant impact on operational efficiency and profitability. Predictive maintenance is a data-driven approach that uses analytics to predict when equipment is likely to fail, allowing businesses to perform maintenance before a breakdown occurs. This proactive approach reduces unplanned downtime, extends the lifespan of equipment, and minimises maintenance costs.

Predictive Maintenance through IoT and Sensors - Predictive maintenance relies on data collected from sensors embedded in machinery and equipment. These sensors monitor various parameters, such as temperature, vibration, pressure, and usage patterns, to detect signs of wear and tear. By analysing this data in real time, businesses can identify potential issues before they lead to equipment failure. For example, in the manufacturing industry, predictive maintenance can be used to monitor the performance of machines on the production line. If the data shows that a machine's vibration levels are increasing beyond a certain threshold, this might indicate that a component is wearing out and needs to be replaced. By addressing the issue before it causes a breakdown, the company can avoid costly downtime and keep production running smoothly. In addition to reducing downtime, predictive maintenance can lead to more efficient use of resources. Traditional maintenance practices often rely on fixed schedules equipment is serviced at regular intervals regardless of whether it needs maintenance. Predictive maintenance, on the other hand, ensures that maintenance is performed only when necessary, reducing the cost of unnecessary repairs and extending the lifespan of equipment.

Asset Optimisation - Beyond predictive maintenance, analytics can also be used for asset optimisation the process of maximising the performance and value of a company's physical assets. This involves using data to monitor the usage, condition, and performance of assets such as machinery, vehicles, or infrastructure. For instance, in the transportation industry, fleet management companies use analytics to track the performance of their vehicles. By analysing data on fuel consumption, driver behaviour, and maintenance history, companies can optimise their fleet operations to reduce fuel costs, extend the lifespan of vehicles, and improve overall efficiency. This level of asset optimisation helps companies operate more sustainably and cost-effectively, giving them a competitive advantage in the market.

Enhancing Workforce Productivity through Data-Driven Insights

In addition to optimising processes and equipment, data analytics can be used to enhance workforce productivity. By analysing data on employee performance, engagement, and workload distribution, businesses can identify opportunities to improve productivity, allocate resources more effectively, and create a more efficient workforce.

Workforce Performance Analytics - One-way businesses can use data to improve workforce productivity is by analysing employee performance. This involves tracking key metrics such as output, efficiency, accuracy, and task completion times. By identifying patterns in this data, businesses can gain insights into which employees are performing at the highest levels, and which may need additional support or training. For example, in a sales organisation, analytics might reveal that certain salespeople consistently outperform their peers in terms of closing deals or meeting targets. By understanding what these top performers are doing differently such as their approach to prospecting, relationship building, or time management the company can replicate their best practices across the broader sales team. Similarly, in a call centre environment, data analytics can be used to monitor call handling times, customer satisfaction scores, and resolution rates. By identifying the characteristics of high-performing agents, businesses can provide targeted coaching or training to help underperforming agents improve.

Workforce Optimisation and Scheduling - Data analytics can also be used to optimise workforce scheduling and resource allocation. By analysing data on demand patterns, workload distribution, and employee availability, businesses can ensure that they have the right number of employees in place at the right times. For example, in the retail industry, analytics can be used to optimise staff scheduling based on foot traffic data and sales patterns. If the data shows that customer traffic is highest during certain hours or days of the week, the company can schedule more employees during those times to ensure that customer needs are met. This level of optimisation not only improves customer service but also reduces labour costs by ensuring that employees are only scheduled when they are needed.

Real-World Examples of Operational Innovation through Analytics

To illustrate how businesses can use analytics to drive operational innovation, let's look at a few real-world examples:

General Electric: Predictive Maintenance in Industrial Equipment - General Electric (GE) has pioneered the use of predictive maintenance in the industrial sector through its Predix platform. Predix collects data from sensors embedded in industrial equipment such as turbines, jet engines, and power plants, allowing GE to monitor the performance of these assets in real-time. By analysing this data, GE can predict when equipment is likely to fail and schedule maintenance before a breakdown occurs. This proactive approach to maintenance has significantly reduced downtime for GE's customers, improving operational efficiency and extending the lifespan of critical equipment. By using data to optimise maintenance schedules and prevent unexpected failures, GE has been able to deliver greater value to its customers while reducing costs.

UPS: Route Optimisation through Predictive Analytics - UPS has long been a leader in using data analytics to optimise its delivery operations. Through its ORION (On-Road Integrated Optimisation and Navigation) system, UPS uses predictive analytics to determine the most efficient delivery routes for its drivers. ORION considers factors such as traffic conditions, weather patterns, and package volume to recommend the optimal route for each delivery. By optimising delivery routes, UPS has been able to reduce fuel consumption, lower emissions, and improve delivery times. This level of operational efficiency has given UPS a significant competitive advantage in the logistics industry, allowing the company to offer faster and more reliable delivery services than its competitors.

Procter & Gamble: Supply Chain Optimisation through Data Analytics - Procter & Gamble (P&G) has used data analytics to optimise its supply chain operations, reducing costs and improving efficiency. By analysing data on inventory levels, supplier performance, and demand forecasts, P&G has been able to streamline its procurement and production processes. For example, the company uses predictive analytics to forecast demand for specific products, allowing it to adjust production schedules and inventory levels in real time. This data-driven approach to supply chain management has enabled P&G to reduce excess inventory, minimise stockouts, and improve overall operational performance. By using analytics to optimise its supply chain, P&G has been able to respond more effectively to changes in customer demand and market conditions.

Implementing Data-Driven Operational Innovation

For businesses looking to drive operational innovation through analytics, the following steps can help ensure success:

Collect Comprehensive Data: The first step in operational innovation is collecting comprehensive data on every aspect of your operations, including process performance, equipment usage, workforce productivity, and supplier performance.

Invest in Analytics Tools: To analyse and interpret the data, businesses need to invest in advanced analytics tools, such as predictive analytics platforms, business intelligence (BI) software, and machine learning algorithms.

Monitor and Iterate: Operational innovation is an ongoing process. Businesses should continuously monitor their processes, equipment, and workforce performance to identify areas for improvement and make data-driven adjustments.

Foster a Data-Driven Culture: Successful operational innovation requires a culture where data is valued and used to inform decision-making at all levels of the organisation. Innovation in operations and processes through analytics provides businesses with the insights needed to optimise workflows, reduce costs, improve efficiency, and enhance overall performance. By leveraging data to streamline processes, optimise supply chains, and enhance workforce productivity, businesses can create a more agile and competitive operation. In the next chapter, we will explore how businesses can use analytics to drive continuous improvement and foster a culture of innovation that spans every aspect of the organisation.

9

DRIVING CONTINUOUS IMPROVEMENT THROUGH ANALYTICS.

Continuous improvement is a fundamental principle in business today. It involves the ongoing effort to enhance products, services, operations, and processes, driving higher levels of efficiency, customer satisfaction, and competitiveness. While continuous improvement has long been a staple in various industries, particularly manufacturing, the rise of analytics has taken it to a new level. By providing real-time insights, predictive capabilities, and actionable data, analytics enables organisations to embed continuous improvement into the fabric of their operations and decision-making processes. Rather than relying solely on periodic reviews or occasional audits, businesses that leverage analytics for continuous improvement can continuously monitor their performance, identify bottlenecks, and implement data-driven enhancements in real time. This approach ensures that businesses remain agile, responsive to market changes, and capable of delivering sustained value to their customers. In this chapter, we will explore how businesses can use analytics to drive continuous improvement across all aspects of their operations. We'll cover specific techniques such as performance monitoring, root cause analysis, real-time feedback loops, and predictive improvements. Additionally, we'll examine real-world examples of how companies have successfully integrated analytics into their continuous improvement initiatives.

The Role of Analytics in Continuous Improvement

At the heart of continuous improvement is the commitment to making incremental changes that lead to better results over time. Traditionally, businesses relied on manual methods such as audits, reviews, and feedback sessions to identify areas for improvement. While these methods are still valuable, they are often reactive, meaning they address issues after they have already impacted the business. Analytics transforms this approach by enabling businesses to take a proactive stance toward improvement. Through continuous data collection, analysis, and real-time reporting, businesses can monitor their performance on an ongoing basis and make data-driven adjustments as soon as issues arise. This shift from reactive to proactive improvement ensures that businesses can optimise their operations more effectively and remain competitive in a rapidly changing marketplace.

There are several key areas where analytics plays a critical role in continuous improvement:

Performance Monitoring: Tracking key performance indicators (KPIs) and metrics to identify areas of improvement in real time.

Root Cause Analysis: Using data to identify the underlying causes of issues or inefficiencies.

Real-Time Feedback: Implementing feedback loops that provide actionable insights on performance and customer satisfaction.

Predictive Improvements: Leveraging predictive analytics to forecast potential issues and address them before they occur.

Performance Monitoring and Key Metrics

Continuous improvement begins with a clear understanding of how the business is currently performing. This requires the consistent monitoring of key performance indicators (KPIs) that are directly linked to the organisation's goals and objectives. KPIs vary depending on the industry and business model, but common examples include metrics related to productivity, quality, customer satisfaction, financial performance, and operational efficiency. Analytics enables businesses to track these KPIs in real time, allowing decision-makers to identify performance gaps as soon as they emerge. Unlike traditional methods that rely on historical data and periodic reviews, analytics platforms such as Power BI, Tableau, or Google Analytics provide real-time dashboards that visualise key metrics in an accessible and actionable format. For example, a manufacturing company might track KPIs related to production output, defect rates, and machine downtime. By monitoring these metrics in real-time, the company can quickly spot when production output is below target, when defects are rising, or when machines are experiencing more downtime than usual. This data allows the business to make immediate adjustments to improve performance before these issues escalate.

Setting Baselines and Targets - For performance monitoring to be effective, businesses must first establish baselines and targets for their key metrics. A baseline represents the current level of performance, while a target represents the desired level of performance. Analytics allows businesses to set these baselines based on historical data and industry benchmarks, providing a reference point for continuous improvement efforts. Once baselines and targets are established, businesses can use analytics to track their progress toward achieving these goals. For example, a retail company might set a target to reduce customer complaints by 10% over the next quarter. By monitoring complaint data in real-time and comparing it to the baseline, the company can evaluate whether it is on track to meet its target and take corrective action if necessary.

Root Cause Analysis through Data - While performance monitoring helps businesses identify when and where issues are occurring, root cause analysis (RCA) is essential for understanding why these issues are happening. Analytics provides the tools needed to perform comprehensive root cause analysis by examining data from multiple sources and identifying the underlying factors that contribute to performance problems. For example, if a company notices that its production output has decreased, it might use root cause analysis to determine whether the issue is due to machine malfunctions, supply chain disruptions, or staffing shortages. By analysing data from production logs, supplier reports, and workforce schedules, the company can pinpoint the exact cause of the issue and implement targeted solutions to address it. In customer service, root cause analysis can be used to identify the reasons behind a decline in customer satisfaction. For instance, if customer feedback indicates that response times have increased, a business might analyse call centre data to determine whether the issue is related to staffing levels, agent performance, or system inefficiencies. By identifying the root cause, the business can make data-driven improvements to enhance the customer experience.

The 5 Whys Technique

One common method used in root cause analysis is the 5 Whys technique, which involves asking "why" five times (or more) to drill down into the underlying causes of a problem. Analytics enhances this technique by providing data at each stage of the questioning process, allowing businesses to uncover more precise and actionable insights.

For example, let's say a company's product defect rate has increased:

Why did the defect rate increase?

Because the assembly line is producing faulty components.

Why is the assembly line producing faulty components?

Because the machine calibration is off.

Why is the machine calibration off?

Because regular maintenance has not been performed.

Why was regular maintenance not performed?

Because the maintenance schedule was delayed.

Why was the maintenance schedule delayed?

Because there was a shortage of maintenance staff.

Through this process, the company can identify that the root cause of the defects is a staffing issue in the maintenance department. By using data from maintenance logs and staffing reports, the company can develop a solution to address this root cause, such as hiring additional maintenance personnel or implementing predictive maintenance practices.

Real-Time Feedback Loops for Continuous Improvement

Incorporating real-time feedback loops into the continuous improvement process is another powerful way to leverage analytics. Real-time feedback loops involve the collection of data on an ongoing basis, providing immediate insights into how a product, service, or process is performing. This feedback is then used to make incremental adjustments that drive continuous improvement. For example, in the software industry, companies often use real-time feedback loops to monitor how users interact with their applications. By collecting data on user behaviour, such as how long they spend on specific features, which areas they click on most frequently, or where they encounter difficulties, software developers can make data-driven improvements to enhance the user experience. In customer service, real-time feedback loops can be used to monitor customer satisfaction in real-time. Many businesses implement post-interaction surveys that ask customers to rate their experience immediately after engaging with a company's service team. By analysing this data in real time, businesses can identify patterns of dissatisfaction and take immediate action to improve the customer experience.

Agile and Iterative Improvement

Real-time feedback loops are particularly valuable in environments where agile and iterative improvement are central to the business model. Agile methodologies, commonly used in software development, involve releasing incremental updates to a product or service and continuously improving it based on user feedback and performance data. For example, a tech company might release a new feature in its app and then use real-time feedback to monitor how users engage with the feature. If the data shows that users are struggling with certain aspects of the feature, the company can make immediate adjustments, such as redesigning the user interface or providing additional instructions. This iterative approach ensures that the product is continuously improved based on actual user feedback rather than assumptions.

Predictive Analytics for Proactive Improvements

While real-time feedback and root cause analysis are essential for addressing current issues, predictive analytics takes continuous improvement a step further by forecasting potential issues and identifying opportunities for proactive improvement. Predictive analytics uses historical data, machine learning algorithms, and statistical models to predict future outcomes, enabling businesses to make improvements before problems arise. For example, in the manufacturing industry, predictive analytics can be used to forecast when machines are likely to fail based on data from sensors and maintenance logs. By predicting when maintenance will be needed, businesses can schedule repairs in advance, preventing costly downtime and ensuring that production remains on track. In customer service, predictive analytics can be used to anticipate customer complaints or service disruptions based on past data. For instance, if a company knows that customer demand tends to spike during certain times of the year, it can use predictive analytics to forecast these spikes and allocate additional resources to handle the increased volume. By proactively addressing these issues, businesses can avoid service disruptions and maintain high levels of customer satisfaction.

Predictive Improvements in Supply Chain Management

Supply chain management is another area where predictive analytics can drive continuous improvement. By analysing data on supplier performance, inventory levels, and transportation costs, businesses can predict potential disruptions in the supply chain and take proactive measures to mitigate them. For example, a retailer might use predictive analytics to forecast when certain products are likely to experience stockouts based on historical sales data, shipping delays, and supplier performance. By anticipating these stockouts, the retailer can place orders in advance, adjust its inventory levels, or explore alternative suppliers to ensure that it can meet customer demand without interruption.

Continuous Improvement through Analytics

Let's look at a few real-world examples of how companies have used analytics to drive continuous improvement:

Toyota: Lean Manufacturing and Real-Time Data - Toyota is widely regarded as a pioneer in continuous improvement, thanks to its adoption of the Lean manufacturing methodology. One of the core principles of Lean is the concept of Kaisen, which emphasises continuous, incremental improvements to processes. Toyota uses real-time data from its production lines to monitor performance and identify areas for improvement on an ongoing basis. By tracking metrics such as production speed, defect rates, and machine downtime in real time, Toyota can quickly identify inefficiencies and implement improvements. For example, if a machine's performance begins to decline, Toyota's data analytics system alerts the maintenance team, allowing them to perform repairs before the machine fails. This approach has helped Toyota maintain its reputation for high-quality, efficient manufacturing.

Amazon: Continuous Optimisation of Logistics and Operations - Amazon's success is largely built on its ability to optimise its logistics and operations through data-driven continuous improvement. Amazon uses real-time data to monitor every aspect of its supply chain, from inventory levels to delivery times. By continuously analysing this data, Amazon can identify inefficiencies, such as slow shipping routes or bottlenecks in its warehouses and implement improvements to enhance its operations. For example, Amazon uses predictive analytics to optimise its inventory management, ensuring that products are always in stock and delivered quickly to customers. By forecasting demand and adjusting its inventory levels in real time, Amazon can reduce the risk of stockouts and improve customer satisfaction.

Netflix: Continuous Improvement of User Experience through Data Netflix uses data analytics to continuously improve its user experience by analysing how viewers engage with its platform. By collecting data on viewing habits, search patterns, and user preferences, Netflix can make data-driven improvements to its recommendation algorithm, ensuring that users are presented with content that aligns with their interests. Netflix also uses real-time feedback loops to monitor how users respond to new features or interface updates. If the data shows that users are struggling to navigate a new feature, Netflix can quickly adjust improve usability. This iterative approach to improvement has helped Netflix maintain its position as a leader in the streaming industry.

Implementing Analytics-Driven Continuous Improvement

For businesses looking to implement analytics-driven continuous improvement, the following steps can help ensure success:

Identify Key Metrics: Start by identifying the key metrics that are most closely tied to your business objectives. These metrics will form the foundation of your continuous improvement efforts.

Invest in Real-Time Analytics: Implement real-time analytics tools that allow you to monitor performance continuously and make data-driven adjustments as soon as issues arise.

Establish Feedback Loops: Create real-time feedback loops that provide actionable insights from customers, employees, and operational processes. Use this feedback to make incremental improvements.

Use Predictive Analytics for Proactive Improvement: Leverage predictive analytics to forecast potential issues and opportunities for improvement before they occur. This allows you to stay ahead of problems and continuously optimise your operations. Analytics provides businesses with the tools and insights needed to drive continuous improvement across every aspect of their operations. From real-time performance monitoring to predictive improvements, analytics enables businesses to identify issues before they escalate, optimise processes on an ongoing basis, and stay agile in an ever-changing market. In the next chapter, we will explore how businesses can use analytics to foster a culture of innovation and ensure that their continuous improvement efforts are sustainable over the long term.

10

FOSTERING A SUSTAINABLE CULTURE OF INNOVATION WITH ANALYTICS.

In today's fast-paced, competitive business environment, innovation cannot be a one-time event or limited to isolated departments. To thrive over the long term, organisations must foster a sustainable culture of innovation one in which continuous improvement, creativity, and experimentation are embedded into the very fabric of the business. Central to this shift is the power of data analytics. By leveraging analytics, businesses can unlock the insights needed to drive ongoing innovation, make smarter decisions, and adapt to changing market conditions. Fostering a culture of innovation requires more than just technological tools; it demands a transformation in the way organisations think, collaborate, and make decisions. Analytics plays a crucial role in this transformation by enabling data-driven decision-making, democratising access to information, and creating a feedback loop that allows businesses to constantly refine and optimise their strategies. In this chapter, we will explore how organisations can use analytics to build and sustain a culture of innovation that permeates every level of the business. We'll cover the importance of leadership, collaboration, experimentation, and accountability, as well as provide real-world examples of companies that have successfully created innovative cultures using analytics.

The Role of Leadership in Building a Culture of Innovation

Innovation starts at the top. For an organisation to foster a sustainable culture of innovation, its leaders must champion the use of analytics and encourage a mindset that embraces experimentation, learning from failure, and constant adaptation. Leadership plays a key role in setting the tone for how data is used across the organisation, creating an environment where employees are empowered to take risks and innovate without fear of failure.

Data-Driven Leadership - Data-driven leadership is critical for embedding analytics into the decision-making processes of the organisation. When leaders prioritise the use of data in their own decision-making, they signal to the rest of the organisation that analytics is not just a tool but a strategic asset. Data-driven leaders rely on insights from analytics to inform their choices, whether it's launching new products, entering new markets, or optimising operations. For example, a CEO who regularly consults data on customer satisfaction, employee engagement, and financial performance before making strategic decisions sets a powerful example for the rest of the organisation. This approach encourages other leaders and teams to adopt similar practices, integrating data into their daily workflows and innovation processes.

Encouraging Risk-Taking and Experimentation - Leaders also play a crucial role in fostering a culture where risk-taking and experimentation are encouraged. Innovation inherently involves uncertainty, and employees must feel empowered to explore new ideas without fear of failure. When leaders promote a growth mindset, they create a safe environment where teams can experiment with new products, services, or processes, knowing that failures are simply opportunities to learn and improve. To encourage risk-taking, leaders can establish innovation incubators or dedicated teams responsible for exploring new ideas and testing them through pilot programs. By providing these teams with the resources, data, and autonomy they need to innovate, leaders help ensure that innovation becomes a continuous, organisation-wide effort rather than a one-time project.

Democratising Data Access to Drive Innovation - For innovation to be sustainable, data must be accessible to everyone in the organisation, not just to those in specialised roles or leadership positions. Democratising data access means providing employees at all levels with the tools and information they need to make data-driven decisions, experiment with new ideas, and contribute to the innovation process.

Self-Service Analytics Tools - The rise of self-service analytics tools such as Tableau, Power BI, and Google Data Studio has made it easier than ever for non-technical employees to access and analyse data without relying on IT or data science teams. These tools enable employees to generate reports, create visualisations, and gain insights from data independently, empowering them to use data in their daily decision-making. For example, a marketing team can use self-service analytics tools to track the performance of different campaigns in real time, adjusting their strategies based on data insights. A product development team can use data to analyse customer feedback and prioritise features for the next product iteration. By putting data in the hands of employees, businesses enable more agile, informed, and creative decision-making across the organisation.

Fostering Data Literacy - While democratising access to data is essential, it must be accompanied by efforts to foster data literacy. Data literacy refers to the ability of employees to understand, interpret, and use data effectively. Without data literacy, even the most advanced analytics tools will be underutilised. To foster data literacy, businesses can implement training programs that teach employees how to work with data, from basic concepts such as understanding metrics and visualisations to more advanced techniques such as statistical analysis or predictive modelling. These programs should be tailored to the specific needs and skill levels of different teams, ensuring that everyone in the organisation has the skills they need to leverage data in their roles. Organisations can also foster data literacy by creating a culture where data is valued and discussed openly. Regularly sharing data insights across teams, celebrating data-driven successes, and encouraging collaboration between data experts and other departments can help create a culture where data is not just seen as a technical resource but as a vital part of the innovation process.

Creating a Feedback Loop for Continuous Innovation

A key component of building a culture of innovation is creating a feedback loop that enables continuous learning, improvement, and adaptation. Analytics plays a critical role in this feedback loop by providing real-time insights into the performance of products, services, and processes, as well as customer and employee satisfaction.

Real-Time Data and Iterative Innovation - One of the most powerful aspects of analytics is its ability to provide real-time data that can be used to drive iterative innovation. Rather than waiting for quarterly reports or annual reviews, businesses can use real-time data to monitor the impact of their innovations on an ongoing basis and adjust as needed. For example, a software company might use real-time data on user interactions to track the performance of a new feature immediately after it is released. If the data shows that users are struggling to navigate the feature or that it is not being used as expected, the company can make iterative improvements based on this feedback. This iterative approach to innovation allows businesses to refine their offerings continuously, ensuring that they remain aligned with customer needs and preferences. In customer service, real-time feedback can be used to monitor customer satisfaction and resolve issues quickly. By analysing data from customer interactions, surveys, or social media, businesses can identify areas where service quality is falling short and implement improvements in real time.

Customer-Centric Innovation - Another important aspect of the feedback loop is customer-centric innovation the practice of using customer feedback and data to guide the development of new products, services, or features. By analysing customer behaviour, preferences, and feedback, businesses can gain valuable insights into what customers truly want and use this information to drive innovation. For example, e-commerce companies often analyse data from customer reviews, purchase histories, and browsing patterns to identify trends and preferences. By understanding which products are most popular, which features are in high demand, and which pain points customers are experiencing, businesses can develop new offerings that better meet customer needs. Customer-centric innovation can also extend to service industries. For instance, a hotel chain might use data from guest feedback to identify areas where the guest experience can be improved, such as by offering personalised services or upgrading amenities. By continuously gathering and analysing customer data, businesses can ensure that their innovations are not only creative but also relevant and impactful.

Encouraging Collaboration Across Teams

Innovation does not happen in silos. To build a sustainable culture of innovation, businesses must encourage cross-functional collaboration that brings together diverse perspectives, skills, and expertise. Analytics can facilitate collaboration by providing a common language and framework for decision-making, allowing teams from different departments to work together more effectively.

Data-Driven Collaboration - Data provides a shared foundation for collaboration across teams. Whether it's marketing, product development, operations, or customer service, analytics allows different teams to align their efforts around common goals and metrics. For example, a marketing team might collaborate with the sales team to analyse customer data and develop targeted campaigns that resonate with specific customer segments. By using data to inform their decisions, teams can work together to create more effective strategies and drive innovation. Collaboration can also extend beyond internal teams to include external partners, such as suppliers, distributors, or even customers. By sharing data and insights with these partners, businesses can co-create new products, services, or processes that benefit everyone involved.

Breaking Down Silos - To foster collaboration, businesses must actively work to break down silos that separate different departments or teams. Siloed organisations often struggle with communication, coordination, and innovation, as teams are disconnected from each other and fail to share knowledge or insights. Analytics can help break down these silos by providing a centralised platform where data is accessible to everyone, regardless of their role or department. For example, a retail company might use a shared analytics platform that allows different teams such as supply chain, marketing, and customer service to access the same data on inventory levels, sales trends, and customer feedback. By working from the same set of insights, these teams can collaborate more effectively to solve problems, streamline operations, and develop new offerings.

Accountability and Metrics for Sustained Innovation

Finally, building a sustainable culture of innovation requires accountability. Innovation initiatives must be measured, monitored, and evaluated to ensure that they are delivering results. Analytics provides the tools needed to track the performance of innovation efforts and hold teams accountable for achieving their goals.

Setting Innovation KPIs - To ensure that innovation efforts are aligned with business objectives, businesses should establish key performance indicators (KPIs) that measure the success of their innovation initiatives. These KPIs can vary depending on the organisation's goals but might include metrics such as the number of new products launched, the percentage of revenue generated from new products, customer satisfaction scores, or process efficiency improvements. By tracking these KPIs through analytics, businesses can evaluate the effectiveness of their innovation strategies and make data-driven decisions about where to invest their resources. For example, if a company's KPI is to increase the percentage of revenue generated from new products, it can use analytics to track the sales performance of its recent launches and identify which products are contributing the most to revenue growth.

Rewarding Innovation - In addition to measuring performance, businesses should also create incentives that reward innovation. Recognising and rewarding employees or teams that contribute to innovation helps reinforce the importance of innovation and motivates others to participate. Rewards can take many forms, from financial incentives to public recognition or opportunities for career advancement. For example, a company might establish an innovation award that recognises employees who have successfully used data to drive creative solutions, launch new products, or improve processes. By celebrating innovation and making it a core part of the company's culture, businesses can encourage employees to take risks, experiment with new ideas, and continuously contribute to the company's growth.

Real-World Examples of Building a Culture of Innovation Through Analytics

Google: Fostering Innovation through Data-Driven Leadership - Google is widely known for its data-driven approach to innovation. The company's leadership encourages a culture of experimentation and data-driven decision-making, providing employees with the resources and autonomy to innovate. Google uses data to track the performance of its products and services in real time, allowing the company to make continuous improvements based on user feedback. One example of this approach is Google's use of A/B testing, where different versions of a product or feature are tested with users to determine which performs better. By using data to inform these experiments, Google ensures that its innovations are driven by user preferences and behaviours.

Amazon: Innovation through Customer-Centric Analytics - Amazon has built its entire business model around customer-centric innovation, using data to understand and anticipate customer needs. The company's recommendation engine, for instance, analyses data on customer browsing and purchasing behaviour to offer personalised product suggestions. This data-driven approach to innovation has helped Amazon maintain its leadership in the eCommerce industry. In addition to personalising the customer experience, Amazon uses data to optimise its operations, from inventory management to logistics. By continuously gathering and analysing data on customer behaviour and operational performance, Amazon can drive ongoing innovation in both its products and services.

3M: Embedding Innovation into the Culture through Collaboration - 3M is known for its culture of innovation, which is driven by cross-functional collaboration and data-driven decision-making. The company encourages employees to spend 15% of their time working on projects outside of their immediate responsibilities, fostering an environment where new ideas can emerge. By using data to inform their experimentation and development processes, 3M employees can test new ideas, share insights across teams, and collaborate on innovations that span multiple departments. This collaborative, data-driven approach has led to the creation of iconic products like Post-it Notes and Scotch Tape.

Implementing a Sustainable Culture of Innovation with Analytics

For businesses looking to foster a sustainable culture of innovation through analytics, the following steps can help ensure success:

Lead by Example: Ensure that leaders champion data-driven decision-making and encourage risk-taking and experimentation across the organisation.

Democratise Data Access: Provide employees at all levels with the tools and information they need to make data-driven decisions and contribute to innovation.

Create Feedback Loops: Implement real-time feedback loops that allow teams to continuously monitor performance and make iterative improvements.

Encourage Collaboration: Break down silos and promote cross-functional collaboration by providing a centralised platform for data sharing and decision-making.

Measure and Reward Innovation: Establish KPIs for innovation and reward teams that successfully use data to drive creative solutions and improvements.

Fostering a sustainable culture of innovation requires a combination of data-driven leadership, democratised access to data, real-time feedback loops, and cross-functional collaboration. By embedding analytics into every aspect of the organisation, businesses can ensure that innovation is continuous, scalable, and aligned with their long-term goals. As we conclude this book, the future of innovation lies in the strategic use of data empowering businesses to make smarter decisions, drive creativity, and sustain competitive advantage in an ever-evolving world.

GLOSSARY

A/B Testing

A method of comparing two versions of a product, service, or process to determine which one performs better. It involves splitting a sample group into two segments and exposing each group to a different version, then analysing the results to identify which version yields better outcomes.

Agile Methodology

A project management and product development approach that emphasises flexibility, collaboration, and iterative progress. Agile methodologies focus on delivering small, incremental improvements rather than a single, large project completion.

Analytics

The systematic computational analysis of data or statistics, used to uncover patterns, relationships, and insights that inform decision-making, problem-solving, and strategic planning.

Artificial Intelligence (AI)

A branch of computer science that focuses on the creation of intelligent machines capable of performing tasks that typically require human intelligence, such as visual perception, speech recognition, decision-making, and language translation.

Baseline

The initial set of data or performance metrics that serves as a starting point for measuring progress. Baselines are used in performance monitoring to track improvements over time.

Big Data

Extremely large datasets that are too complex to be processed by traditional data-processing software. Big Data analytics refers to the examination of these large datasets to discover hidden patterns, correlations, and other insights.

Business Intelligence (BI)

Technologies, applications, and practices used to collect, integrate, analyse, and present business data to support better decision-making. BI tools and systems help organisations understand their data and make informed decisions.

Collaboration

The act of working together across teams or departments to achieve shared goals. Cross-functional collaboration in innovation often involves diverse groups bringing their skills and perspectives together to solve problems or create new products and services.

Continuous Improvement

An ongoing effort to improve products, services, or processes by making incremental improvements over time. Analytics can drive continuous improvement by providing data-driven insights for process optimisation and operational efficiency.

Customer-Centric Innovation

An approach to innovation that focuses on understanding and meeting the needs of the customer. This involves using customer feedback and data to guide product and service development, ensuring that new offerings align with customer preferences.

Data-Driven Decision-Making

The practice of making decisions based on data analysis and interpretation, rather than intuition or personal experience. Data-driven decision-making involves using real-time or historical data to guide strategic and operational choices.

Data Literacy

The ability to read, understand, and communicate data. Data literacy is critical for employees to effectively use analytics tools, interpret insights, and make data-driven decisions.

Democratising Data

Making data accessible to everyone in the organisation, regardless of their role or technical expertise. Democratising data enables employees at all levels to use analytics to drive decision-making and innovation.

Dynamic Pricing

A pricing strategy that involves adjusting prices in response to real-time demand, competition, or market conditions. Dynamic pricing allows businesses to optimise revenue by offering the right price at the right time.

Feedback Loop

A process in which data is continuously collected, analysed, and used to make iterative improvements. Feedback loops provide real-time insights into performance and help organisations refine their strategies and innovations.

Innovation Incubator

A dedicated team or space within an organisation where employees can explore new ideas, experiment, and develop innovations. Innovation incubators provide resources and support for employees to test and launch new products or processes.

Internet of Things (IoT)

A network of physical devices, vehicles, appliances, and other objects embedded with sensors and software, enabling them to connect and exchange data. IoT devices collect data that can be used for analytics and predictive maintenance.

Key Performance Indicator (KPI)

A measurable value that indicates how effectively a company is achieving its key business objectives. KPIs are used to evaluate the success of an organisation or a particular activity in meeting its goals.

Lean Methodology

A methodology that focuses on reducing waste, improving quality, and optimising efficiency. Lean principles are often applied in manufacturing but can be used in any industry to streamline operations and maximise value.

Machine Learning (ML)

A subset of artificial intelligence (AI) that allows computers to learn from data and make decisions or predictions based on that data. Machine learning algorithms improve over time as they are exposed to more data.

Predictive Analytics

The use of historical data, statistical algorithms, and machine learning techniques to predict future outcomes. Predictive analytics helps businesses anticipate trends, customer behaviours, and operational risks before they occur.

Predictive Maintenance

A data-driven approach to equipment maintenance that uses predictive analytics to forecast when a machine or system is likely to fail. This allows businesses to perform maintenance before breakdowns occur, minimising downtime and repair costs.

Prescriptive Analytics

A type of analytics that uses data to recommend specific actions or solutions to achieve desired outcomes. Prescriptive analytics goes beyond predicting what is likely to happen and suggests the best course of action to take.

Real-Time Data

Data that is collected, processed, and analysed as events occur. Real-time data allows businesses to monitor performance, customer behaviour, and market conditions instantly, enabling faster decision-making and response.

Root Cause Analysis (RCA)

A method of problem-solving that focuses on identifying the underlying cause of an issue. RCA is used to determine why a problem occurred so that it can be addressed and prevented from happening again.

Self-Service Analytics

Analytics platforms that enable non-technical users to access, analyse, and visualise data without requiring assistance from IT or data science teams. Self-service analytics tools make data more accessible across the organisation.

Supply Chain Optimisation

The process of improving the efficiency and performance of a company's supply chain by using data to manage inventory, forecast demand, and streamline logistics. Supply chain optimisation reduces costs, improves delivery times, and enhances customer satisfaction.

Total Quality Management (TQM)

A management approach focused on embedding quality into every aspect of an organisation's operations. TQM emphasises continuous improvement, customer satisfaction, and employee involvement in achieving quality goals.

Workflow Efficiency

The effectiveness with which tasks and processes flow through an organisation. Improving workflow efficiency involves eliminating bottlenecks, reducing redundancies, and optimising resource allocation to maximise productivity.

www.ingramcontent.com/pod-product-compliance
Lightning Source LLC
Chambersburg PA
CBHW021214250926
51852CB00006B/6
* 9 7 8 9 7 9 0 3 5 0 7 6 2 *